The Curious Cookbook

大英图书馆里的秘密食谱

[英] 彼得・罗斯——著　战蓉蓉——译

南方出版传媒
广东经济出版社
- 广州 -

图书在版编目（CIP）数据

大英图书馆里的秘密食谱 / （英）彼得·罗斯著；战蓉蓉译. —广州：广东经济出版社，2020.1
ISBN 978-7-5454-6752-9

Ⅰ.①大… Ⅱ.①彼… ②战… Ⅲ.①食谱-英国 Ⅳ.①TS972.185.61

中国版本图书馆 CIP 数据核字（2019）第131868号

版权登记号：19-2017-144

出 版 人：李 鹏
责任编辑：程梦菲 张晶晶
责任技编：陆俊帆
封面设计：门乃婷

First published by The British Library and Mark Batty Publisher LLC
The simplified Chinese translation rights arranged through Peony Literary Agency

大英图书馆里的秘密食谱
Daying Tushuguan li de Mimi Shipu

出版发行	广东经济出版社（广州市环市东路水荫路11号11~12楼）
经销	全国新华书店
印刷	广东信源彩色印务有限公司（广州市番禺区南村镇南村村东兴工业园）
开本	889毫米×1194毫米 1/32
印张	6.5
字数	170千字
版次	2020年1月第1版
印次	2020年1月第1次
书号	ISBN 978-7-5454-6752-9
定价	59.80元

广东经济出版社官方网站：**http://www.gebook.com** 微博：**http://e.weibo.com/gebook**
图书营销中心地址：广州市环市东路水荫路11号11楼
电话：（020）87393830 邮政编码：510075
如发现印装质量问题，影响阅读，请与承印厂联系调换
广东经济出版社常年法律顾问：胡志海律师

前　言

赫斯顿·布卢门塔尔
(Heston Blumenthal)

法国著名美食家让·安泰尔姆·布里亚-萨瓦兰（Jean Anthelme Brillat-Savarin，1755—1826）曾说过：“从你吃的东西中我就能判断出你是怎样的人。”本书的每一页都能向我们证明在大千世界，在对待吃这件事上，人类敢于尝试（比如吃毒蛇），食材广泛（比如海冬青蜜饯），脑洞大开（比如食用一碟雪花），创意满满（比如用塞满大麦的黄瓜吸引蝇虫），或简单粗糙（比如法式油炸泡芙），或方法奇特（比如带皮烤孔雀），总能寓吃于乐（比如把猪肉、奶酪和面包做成水壶或水罐状），而且无所不吃。这本书囊括了烤水獭、蝰蛇汤、海豚小麦粥、猪脑馅饼、乌龟汤和红烧麻雀吐司等各类千奇百怪的食谱！

我发现探索过去的食谱非常令人着迷，但相对来说这只能算是我刚刚培养起来的一个兴趣爱好。不得不说，十年前我对英国的烹饪历史没有半点想法，这说起来还有点尴尬，开始研究它纯属一个美丽的偶然。我经常冲动性购买一些烹饪书，对买这类书我心怀执念。之前不知从哪里得来一本名叫*The Vivendier*的十五世纪手稿集复印本，其中包含很多食谱，有口喷火焰的鱼，还有看似煮熟的鸡——这种鸡在即将被人切割的时候就会醒来，然后将桌子上可怜的茶壶、酒杯等打翻到地上。

或许是我之前太无知，我从来不知道以前的食谱还可以如此富有创意、一反常态、夸张有趣。为了了解更多，我参加了牛津饮食与烹饪研讨会（Oxford Symposium of Food and Cookery)。该研讨会每年举办一次，讲述一些人们平时了解不到的饮食历史，很有教育意义。那一年的研讨会讲到人类在糟糕处境下可能会食用的东西。我听得头昏脑涨，随即离开大厅出去透了透气，最后和两位男士闲聊起来。这两位男士不仅也知道*The Vivendier*这本书，还了解很多过去的奇异菜肴，比如状如瓶子的鸡。

我认为在牛津的这个烹饪研讨会上，我能够恰巧在两位来自汉普顿宫的食物历史学家身边“透气”，绝非偶然。后来通过他们，我认识了其他食物历史学家并开始阅读一些著作，比如《食物准备法》（*The Forme of Cury*，1390）和《简易烹饪艺术》（*The Art of Cookery Made Plain and Easy*），当然我也看了由无敌的罗伯特·梅（Robert May）写的《成就厨师》（*The Accomplish't cook*)。罗伯特·梅让人难忘的代表食谱有裹着活青蛙和小鸟的馅饼，沾着血的鹿肉糕点，爆炸的船只和城堡，炖牛颚、牛唇和牛鼻。这些书让我们了解到很多难以想象的饮食和烹饪发展史。比如，在《食物准备法》一书中就有藏红花米饭的食谱，让我们了解到这一菜品出现在米兰烩饭之前。在《简易烹饪艺术》（*The Art of Cookery Made Plain and Easy*，1747）中，汉娜·格拉斯(Hannah Glasse）提到了汉堡香肠的制作，这说明汉堡包早在十八世纪就已经出现了。当我对以前的食谱有自己的认识的时候，会经常向它们请教，比如肉味水果、牛肉冻、白杨布丁和醉蛋糕等。

当然，对一个厨师来说，研究古老的食谱然后将它们进行调整改良以满足现代人的口味是一件令人兴奋的事情。这使得我们能够与过去两相连通，给我们营造了一个传承伟大传统的机会，也为我们了解过去的厨师对食物的想法和做法提供了一种途径。即便他们的烹饪技术还只是处于最初级的阶段，但所展现出来的想象力和创造力却十分振奋人心。

但是我要说的是，《大英图书馆里的秘密食谱》这本书要更富乐趣。吃东西是日常生活中所固有的一部分，所以一个特定时期的食谱能够体现这个时期的很多社会特征（虽然情况通常是这样的：几百年前的烹饪资料大多来自富有和贵族家庭，来自普通家庭的很少）。当你继续往下读时你会发现，在中世纪，菜肴的调料配方是多么富有冒险性。特别是在冬季的那几个月，我们没办法从土地里获取更多的食材，也保证不了土地的产量能够满足我们的需求，而调味品的出现给数量有限的食材创造出了更多的形式和口味，使我们在餐桌上胃口大增。天鹅馅饼食谱的例子表明，调味品的加入不仅让大众接受了食用天鹅肉，而且接纳了将天鹅皮作为时尚饰品！由此，你能真正体会到烹饪并不是一成不变的，烹饪的趋势、需求、资源以及获取新型、刺激食材的方法，永远都在进化发展当中。

所以，无论你是一位想对过去那些非同一般而且盛大的菜肴一显身手的厨师，还是一位研究过去古怪食谱的内行，抑或只是一位对英国错综复杂的烹饪和社会历史怀有兴趣的普通人，这本神奇的书中的每一页都会为你带来愉悦和享受。

引 言

所有人都要吃东西，但是我们平时吃的东西或者说至今为止所吃过的东西都因时因地而异。如今，在某种文化环境中，人们可能喜欢吃狗肉而且并不会为此多愁善感，但是他们可能一想到要吃那种叫作“蓝纹奶酪”的腐坏发酵的牛奶块儿，就会感到恶心。同理，想到先人们吃的那些东西，我们也会有同样的感觉，而且以前的饮食风尚因地域分隔而差别更大。这本书会通过实例向我们展示古今食物之不同，通过了解这些从十五世纪到第二次世界大战期间的上百道食谱，我们就会感受到人类在过去吃的那些东西在现在看来是多么奇怪、有趣、恶心或者仅仅只是单纯地让人好奇。好奇心可能会害死猫，但好奇心却不会找来一道古代英国食谱让我们照着做菜。有了这本书，你就会学到很多譬如孔雀、海豚、鸡冠、云雀、蝰蛇、麻雀、八哥、松鼠尾巴、鲦鱼、乌龟、水獭、天鹅等的烹饪方法。你还会学到如何烤制一磅黄油，如何通过拽拉海鸥的腿来辨别肉质是否新鲜，如何给鳕鱼的鱼鳔填充馅料……

在很久以前印刷术还未出现的时候，中世纪人们所食用的东西是由皇室或贵族家庭手写记录下来的。这些记录通常只是一些重要用餐或大型宴会的食材列表或菜单，其中仅仅包含着一些菜肴的名称和上菜顺序。但是到十四世纪末期，从上层社会家庭中的专用厨师那里产生的食谱在某种程度上被详细地记录了下来。第一本食谱就是在查尔斯二世的命令下写成的，叫作《食物准备法》。这一食谱自然是写了一些王室的膳食，而非那些非常基础的粥、根菜类蔬菜或是那些广大人民大众吃的廉价的鱼或肉。也有一些与食谱相关的十五世纪的文献，比如现今收藏于大英图书馆的哈利父子的手稿，同样记录了上层社会的食谱。在这些非常简短的食谱中，奇怪的配料还有怪异的甜咸搭配经常会勾起我们的好奇心。或许我们一想到要吃孔雀或者海豚就会感到毛骨悚然，但是这些东西在那时却十分昂贵而且只能被上层社会所享用。现在很多人常常会对中世纪皇室喜欢吃的动物的部位嗤之以鼻，比如鸡冠或者牛颚。在这些食谱中用的词语也会让人觉得古怪、可爱，有时还会让人感到震惊。中世纪的一名作家用代词“他”而非“它”来称呼被当作食材的动物，这使得一些早期的食谱显示出了强烈的个人色彩。我们会读到一个食谱教程“取一只孔雀，将他的脖子折断，切断他的喉咙”或者“取一只鼠海豚，像剁鲑鱼那样把他剁碎”。看到这样的食谱，我们在吃东西的时候就会很容易想到这些动物之前还是活蹦乱跳的。也是在中世纪这段时期，我们发现人们喜欢创作一些装饰华丽的食物或菜肴，这些食物不仅能够用来款待宾

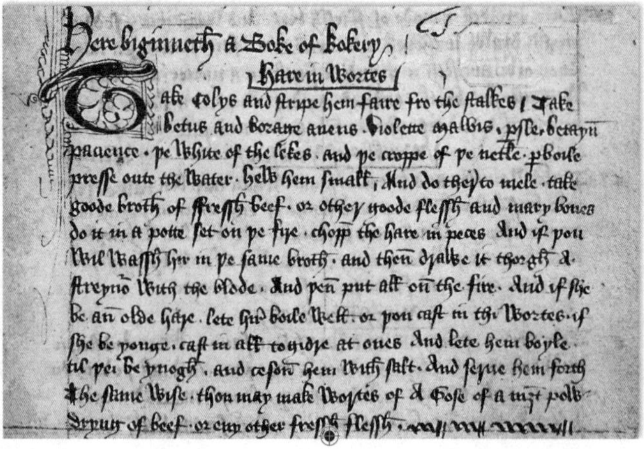
Here biginneth a Boke of Kokery
Hare in Wortes
Take colys and stripe hem faire fro the stalkes. Take
betus and borage auens. violette malvis, persle, betayn
pacience. þe white of the lekes. and þe croppe of þe netle. perboile
presse oute the water. hew hem smale, And do therto mele. take
goode broth of fressh beef. or other goode flessh and mary bones
do it in a potte set on þe fire. chopp the hare in peces And if þou
wil wassh hir in þe same broth. and then drawe it thorgh a
streynour with the blode. And þen put all on the fire. And if she
be an olde hare. lete hir boile well. or þou cast in thi wortes. if
she be yonge. cast in all togidre at ones And lete hem boyle
til þei be ynogh. and seson hem with salt. And serve hem forth
The same wise. thou may make wortes of a Gose of a night polk
Ioynt of beef. or eny other fressh flessh.

客，还可以诱骗和取悦他们。这类食谱有很多，包括：将猪肉、奶酪和面包做成水壶或罐子状、用糖做成小鸟和其他动物的形状，还有在这之后才有的包有活青蛙和小鸟的馅饼。“烤饼里有二十四只画眉”在那时也很容易成为一个真实的食谱名称。

英格兰印刷的第一本食谱是《贵族烹饪法》（*This is the Boke of Cokery*，1500），这是一部稀世珍品，因为对于十六世纪前四分之三的时期来说，食谱大多出现在有关家庭医疗、勤俭持家和牲畜饲养等普通书籍中。然而，在十六世纪末期，很多富有创新精神的伦敦出版商开始收集食谱，出版一些完全关于烹饪的书籍。这其中包括A.W.的《一本烹饪书》（*A Book of Cookrye*，1591），托马斯·道森（Thomas Dawson）的《好主妇是这样的》（*The Good Housewife's Jewel*，1596），休·普拉特（Hugh Plat）编辑出版的《主妇的乐趣》（*Delightes for Ladies*，1602），还有刚进入十七世纪，杰维斯·马卡姆（Gervase Markham）的《英国家庭主妇》（*The English Hus-wife, 1615*）。1615年到1683年间，《英国家庭主妇》这本书再版了13次，这说明烹饪书越来越受到人们的欢迎。1600年到1700年间，几乎每隔一年就会出现一本新的烹饪书，这一数量是十六世纪的两倍还要多。

都铎王朝后期的很多食物和中世纪时期的非常相似：烤制或煮熟的肉、各类汤水（清汤或炖汤）、谷物粥、面包和一些水果蔬菜羹。基本上很少有人会吃生水果或蔬菜，因为大家都觉得吃这些会对身体不好。十六世纪，各式各样的新食物应运而生，包括柑橘、杏子、甜瓜和越来越多的其他水果。它们主要出自新大陆，与非洲、远东日益频繁的贸易往来，使这些新食物开始被越来越多的人接纳、购买。然而在那段时期，人们在饮食上最富戏剧性的变化是对蔗糖的购买需求剧增，至少对那些有钱人来说是这样的。蔗糖从一开始被当作调味品，用量微小，到后来成为甜品、果冻、果酱的主要添加剂，人们甚至开始用纯蔗糖做成糖果。蔗糖的突出地位主要源于那个时期糖果制作这一新门类的出现和发展，起初糖果制作只在有钱人的家里进行，后来独立到商店和包办糖果制作的商人手中。

食材的准备以及配料的使用在十七世纪发生了翻天覆地的变化。那个时候出现了越来越多非贵族出身的有钱人，我们现在可能称其为中产阶级。他们在十七世纪拥有了更多的业余时间并且愿意把钱花在美食和娱乐消遣上。他们当中的很多人都住在城市，所以能够获取各种各样的食材而不必受牵制于是否自产自足。虽然皇室仍然继续主导着饮食的潮流，但这些潮流逐渐自上而下渗透到整个社会。烹饪书不再只是专业宫廷厨师收藏的东西，而开始越来越多地卖给城市家庭中的贵妇们。也正是这些妇女们，在一个世纪之后，创作了她们自己的烹饪书，并且在很长一段时期主导着烹饪书的销售市场。

十七世纪出现了很多新食材，这些新食材当中有一部分成了英国饮食习惯中的主食，主要有茶、咖啡和巧克力。1699年，约翰·伊夫林（John Evelyn）出版了一本叫作《沙拉说》（*Acetaria: A Discourse of Sallets*）的书，整本书专门向人们介绍了生沙拉，直到那时，人们才不再害怕食用生沙拉。此外，土豆、饼干、布丁还有果馅糕点等其他食物也开始受到了人们的欢迎。法国烹饪对英国饮食影响越来越大，特别是在曾被放逐到法国的查尔斯二世复辟之后，再加上十七世纪五十年代拉瓦雷纳（La Varenne）的书也被翻译成了英文，传播到了英国。受到法国烹饪的影响，英国饮食将甜味和咸味分了开来，要知道在这之前甜味和咸味经常被混合在一起，而且在大多数时候都是这样搭配的。十七世纪末，一些像汉娜·伍利（Hannah Wooley）的《贵妇指南》（*The Gentlewoman's Companion*，1673）这种烹饪书，它们的主要销售对象不是农村家庭和宫廷。《贵妇指南》大概是英国第一本由女性创作的烹饪书，几乎只面向城市的中产阶级。另外，被称作是第一批"现代"烹饪书之一的威廉·拉比沙（William Rabisha）的《剖析烹饪全过程》（*The Whole Body of Cookery Dissected*），自1661年出版之后20年内加印了五次，并且直到十八世纪中期一直都保有影响力。当然这也是因为很多饮食习惯在短时间内并没有改变，一顿饭仍旧会有很多道菜，有时会一次性将所有饭菜端到饭桌上，而不是像我们今天这样一道一道上菜。第一道菜可能会将一些食物搭配在一起，通常是汤和烤肉。第一道菜吃完后就会被端走，然后上第二道菜。第二道菜通常是一些开胃菜，包括家禽肉、馅饼和蔬菜，但是也会有一些甜的果馅糕点和水果派。最后一道菜会是一些坚果、新鲜的水果、奶酪和糖果等。

十八世纪早期，写烹饪书的通常是男性，一般是那些曾在贵族家庭里任职的专业厨师，他们写出的食谱能够反映出那些家庭的地位和声望。在这段时期，最有影响力的烹饪书是从法语翻译而来的梅西阿洛特（Massialot）的《宫廷和乡村厨师》（*The Court and Country Cook*，1702），其中介绍了很多复杂的食谱，包括将多种肉放在一起烹饪，然后配上浓酱汁食用，通常还会添加一些别的东西和装饰菜，比如巧克力海鸭子这道菜。很多出版的烹饪书，像伊丽莎·史密斯（Eliza Smith）的《称职的主妇》（*The Compleat Housewife*，1727），就试图模仿或简化梅西阿洛特的烹饪风格。查尔斯·卡特

（Charles Carter）的《卓越的宫廷和乡村厨师们》（*The Compleat City and Country Cook*，1732）在图解城市精英晚餐的餐桌布置的同时，也将一些更划算、更简单的流行食谱包含在其中。相反，牛津的植物学教授理查德·布兰德利（Richard Bradley）对烹饪持有不同的看法，他出版了一本有点与众不同的食谱叫作《乡村主妇》（*The Country Housewife*，1727），这本书收集的食谱全都是他自己感兴趣的，包括本书中提到的蝰蛇汤和獾肉火腿，这显然与当时流行的烹饪风格不符。

十八世纪中期，由女性们写的烹饪书成了烹饪书市场的主导。其中汉娜·格拉斯（Hannah Glasse）的《简易烹饪艺术》（*The Art of Cookery Made Plain and Easy*，1747）在那个时期卖得最好。虽然汉娜·格拉斯声称自己并不喜欢法国烹饪法，她只是将法国烹饪改为适合英国人口味和消费水平的烹饪风格，但是如今汉娜·格拉斯被公认为“借用”了她那个时代下的很多法国食谱。她的初衷是让家庭的女主人能够用这本书指导家中那不中用的厨子。格拉斯在书中提供了制作昂贵酱料的捷径，还提供了用简单的配料装饰菜肴的方法，大家可以看一下奇异鸡蛋和仙女黄油这两道食谱。这种对装饰菜情有独钟的情况在伊丽莎白·拉斐尔德（Elizabeth Raffald）的很多食谱中也有所体现，这些菜一般是摆在盘子中间用于装饰，同时也可以食用，比如拉斐尔德写的嵌在果冻里的母鸡和小鸡。

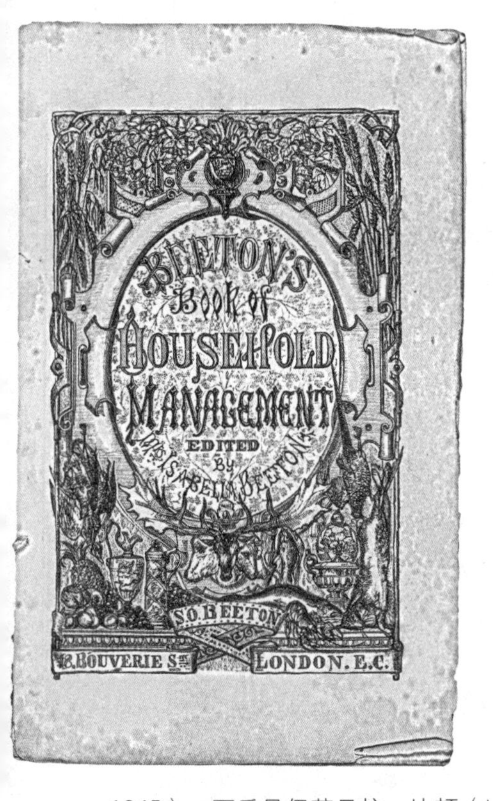

十八世纪末至十九世纪初出现了很多烹饪书，都是由像伦敦酒馆（London Tavern）一样著名酒店里的专业男性厨师们出版的。这种酒店的招牌菜是甲鱼汤，在那个时期甲鱼汤也会出现在许许多多的正式场合甚至是市长大人的宴会上。很多酒店都有养满活甲鱼的地窖，这些甲鱼会被用来加工制作成一桶一桶的甲鱼汤，然后装进罐头里。此外，在那个年代，最畅销的烹饪书仍旧是由女性作家出版的。起初是伊莱扎·阿克顿（Eliza Acton）的《现代家庭烹饪》（*Modern Cookery for Private Families*，1845），而后是伊莎贝拉·比顿（Isabella Beeton）的《家务手册》（*Book of Household Management*，1861）。这两位女性都倾向于探求那种成本非常低但却对身体有益的菜肴，虽然这些菜仍被认为是传统意义上的英国菜，但是她们却能勇敢地将这些菜与皇室流传出来的新想法和配料结合在一起。

二十世纪，很多烹饪书作家开始将比顿女士强调的实惠和快捷看作是英国烹饪逐渐衰退的象征，英国烹饪确实将在二十世纪五十年代早期衰退到最低点。然而这些批评家在肆意批评中却忽略了一点，那就是比顿女士的书远比那些拮据的家庭主妇的烹饪指南要更加全面，其中包括制作复杂的小牛肉、大菱鲆、燕窝汤、果冻和冰激凌等食谱，还包括一些受到欧洲大陆和殖民地影响的烹饪方法。二十一世纪，随着英国烹饪的复兴，比顿女士作为影响深远的烹饪书作家重新获得了公正的评价，尽管事实是她可能像其他烹饪书作家一样“借用”过他人的食谱。

回顾二十世纪，想要发现一些在我们现在看来稀奇古怪的食物或食谱并不容易。就像世界上很多其他的东西一样，人类开始厌倦了不断尝试新事物。但是有一个重要的时间段能给我们提供一些有趣的食谱，并且能将我们带回到那个人人都靠寻觅而来的食物充饥的时期，这个时期就是第二次世界大战期间的“烽火家园”（Home Front）。那时由于食物短缺和定量供应，政府控制着整个国家的粮食，还通过分发小册子和举行活动来鼓舞大家食用那些能够在家里最快生产的东西，就是在这一时期出现了土豆皮特和胡萝卜博士。他们运用不同的烹饪技巧尽可能将食物做得有趣，汇集在最早的名人食谱集之一的《厨房大作战: 定额配给时期150个名人食谱》（*A Kitchen Goes to War：Famous People Contribute 150 Recipes to a Ration-time Cookery Book*，1940）中。

尽管政府做了很多努力，但是土豆、胡萝卜、燕麦以及很少一部分定量下发的“奢侈食品”并做不了太多的东西。一些作家建议用“野外食物”来弥补战时乏味但还算健康的饮食。大概是在中世纪时期，第一次有人提议说人民大众可能会喜欢吃刺猬、秃鼻乌鸦和夜莺。现在很难统计当时有多少人听从了这个建议——其实现在也很难知道有多少人曾经尝试过这本六百年烹饪食谱选集中的任意一个食谱。但也许是因为财富和机遇，又或者是因为贫穷和饥饿，甚至只是因为平常的好奇心，曾让我们的一些祖先在海豚、百灵鸟、乌龟或者一碗热气腾腾的松鼠尾巴汤中摸索前进。

注解：本书采取了“中文翻译食谱+食谱注解+英文食谱”的排版设计，方便读者阅读。在英文食谱中，原始手稿里的标点及拼写绝大多数都被保留，除非它们会严重影响理解。在词语或单词的意思很模糊的情况下，会在方括号里提供它们的翻译或其他注解。因为很多早期的食谱将所有名词的首字母都大写了，这一现象直到十八世纪都很常见，在此将它们修改成现代的形式，仅保留那些专有名词的大写。另外，需要说明的是，在引言中提到的作品出版日期都是首次出版的日期，正文中附在个人食谱后的日期是它们摘自的那一版本的日期。

书中所涉及的计量单位的换算关系如下：

1配克（英）=2加仑（英）=9.0922升；

1加仑（英）=4夸脱（英）=4.54609升；

1夸脱（英）=2品脱（英）=1.1365升；

1品脱（英）=4及耳（英）=5.6826分升；

1及耳（英）=1.4207分升；

1英液盎司=2.8413厘升；

1磅=16盎司=0.4536千克；

1盎司=28.3495克；

1英尺=12英寸=0.3048米；

1英寸=2.54厘米。

另外，书中涉及少量以现代世界珍稀动物为食材的菜谱，还有一些含毒的、不可食用的菜谱，不可轻易模仿哦！

Cucina principale
reduta da pani
lucerna
Camerino per garzoni
ordegno
murello p pignatte
tauola da pasta
bancho
Coloma col mortaro
Tauola per imbandire.

目　录

contents

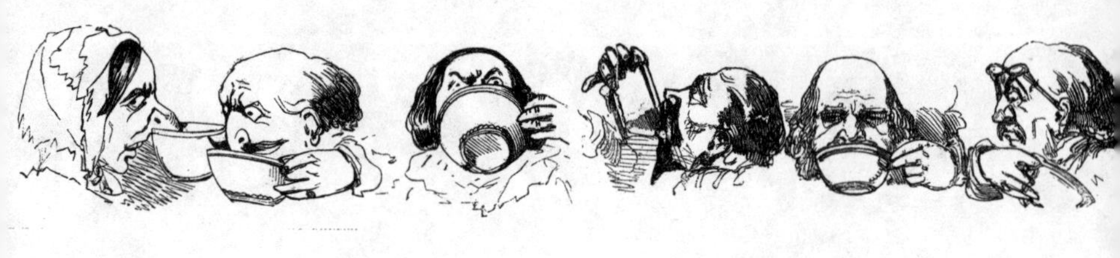

Recipes

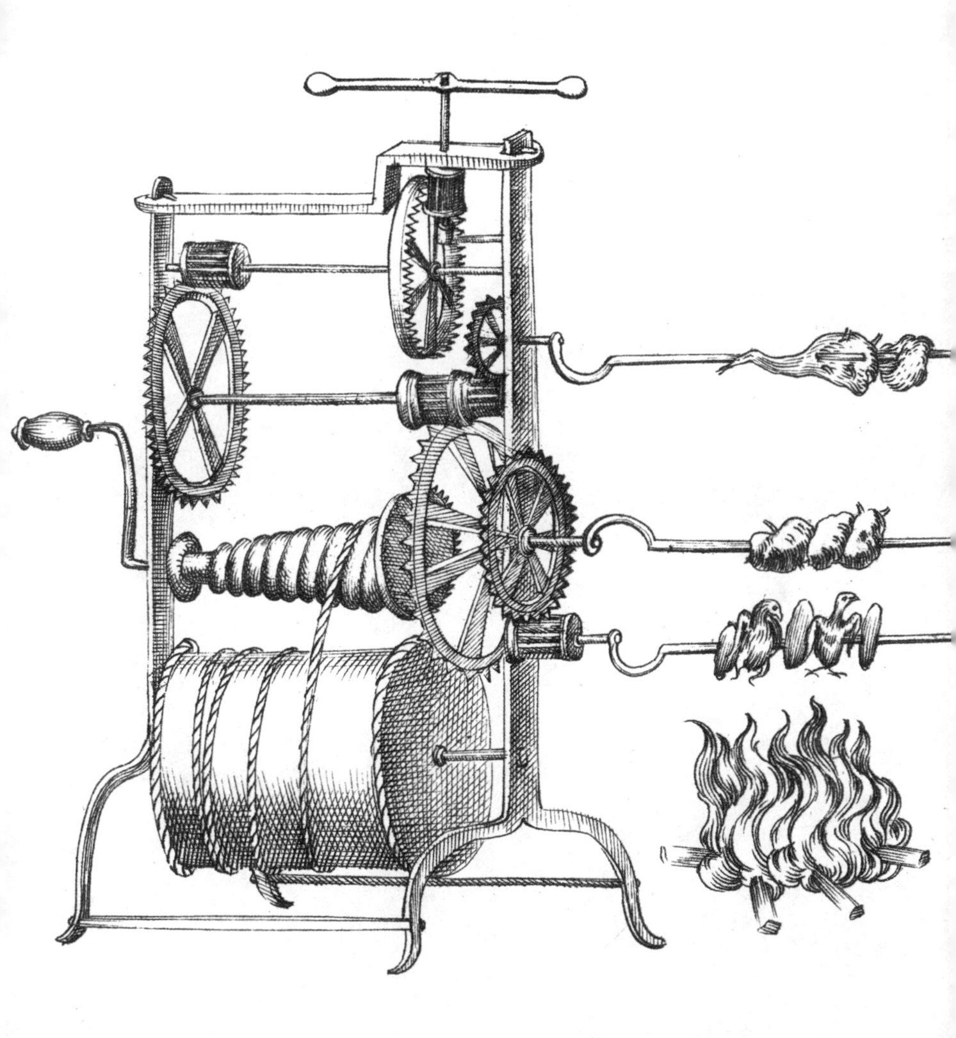

中世纪千层面
A medieval lasagne

取优质清汤，将其倒入砂锅中。再准备一些上好的白面粉，加水揉成面团，然后用擀面杖擀成薄面片。等面片彻底风干后放入清汤中煮熟。在盘子底部撒上磨碎的生奶酪、蔗糖、五香粉，然后尽量将面片完整地放到盘子里。最后，在面片上撒上两倍或三倍量的五香粉和生奶酪，就可以端上桌了。

——《食物准备法》，1390

☘我们现代的千层面显然是从这道美食演化而来的。面条的历史向来饱受争议，据传早在十四世纪，马可·波罗（Marco Polo）从东方游历回来时把面食带回了欧洲。但是，早在马可·波罗出生之前，欧洲就已经有了面条和其他面食，这或许是受到了阿拉伯和罗马饮食的影响。写有这一食谱的那本烹饪书里还有一道制作奶酪馅方形饺子的食谱。

A MEDIEVAL LASAGNE

LOZENGES[1]. *Take good broth and put it in an earthenware pot. Take the best quality white flour and make a paste with water and make thin sheets with a rolling pin. Dry it until hard and boil it in broth. Take raw cheese grated and lay it in a dish with sugar and spice powder, and lay the lozenges as whole as you can, and above that more powder and more cheese, and so twice or thrice, and serve.*

The Forme of Cury, c. 1390

① 英文食谱每段开头的大写英文为原始手稿中的食谱名称。

可食杂菜

Preparing edible 'garbage' and 'compost'

面包鸡杂汤：取新鲜的"食物废料"——鸡头、鸡爪、鸡肝和鸡肫等，将它们洗净，放入装有新鲜牛肉汤的锅中，撒上胡椒粉、桂皮、丁香、肉豆蔻粉、欧芹以及切碎的鼠尾草。另一边，再准备一些牛肉汤，把面包泡在里面，溶化成糊状后过滤一下。把过滤后的面包糊倒入装有"食物废料"的锅里，开始炖煮。煮熟后加入姜末、酸葡萄汁、盐和一点藏红花粉就可以上桌了。

——《哈利父子文稿集》，1450

拌杂蔬：取欧芹根、欧洲萝卜、小萝卜，去皮、洗净。取大头菜和卷心菜，剥净并去掉菜心。在一个陶瓷平底锅里倒入清水，放到火上加热，将所有食材放入其中。煮沸后，放入一些梨块继续煮到半熟，然后将所有食材取出，放到一张干净的布上冷却，撒上盐。当所有食材完全冷却后，将其放到一个器皿里，倒上食醋，撒上调料和藏红花，然后静置一天一夜。之后，准备好希腊红酒、蜂蜜、伦巴第芥末、一整个醋栗、桂皮粉、蔗糖、五香粉和茴香籽，将所有这些配料和前一天的食材搅拌均匀，放到陶制的器皿中，想吃的时候取出来吃就可以了。

——《食物准备法》，1390

♣拌杂蔬是英国人喜欢吃的一种调味菜或者说是咸菜，与此类似的食谱可以追溯到罗马美食家阿比修斯（Apicius）所写的烹饪书中。阿比修斯在公元一世纪就写了一本烹饪书，他书中记载的调味菜发展为现代很多食谱中的酸辣酱。拌杂蔬和辣泡菜也十分相似。辣泡菜是由英国的一种蔬菜做成的咸菜，与意大利的一道叫作芥末水果蜜饯的食谱有很多共同之处。

♣“食物废料（garbage）”一词源于英语，用于形容那些在厨房中会被丢弃的一部分食物，特别是家禽的内脏杂碎——肫、肝脏、心脏，还有家禽的脖子等。很难相信在中世纪很多家庭都会把鸡的很多明明可以食用的部位丢掉。有趣的是，中世纪筛选香料时也要将不好的香料从好香料中择选出来。“筛选（garble）”一词好像与此有紧密的联系，它最初的含义是指“将最好的从……中取出”，现在的含义则与之完全相反——把东西混合或混淆在一起。

PREPARING EDIBLE 'GARBAGE' AND 'COMPOST'

GARBAGE. *Take fresh garbage, chickens' heads, feet, livers and gizzards, and wash them clean. Cast them into a good pot with fresh beef broth, powdered pepper, cinnamon, cloves, mace, parsley and sage minced small. Then take bread, soak it in the same broth, push it through a strainer and add it to the pot. Let it boil sufficiently and then add powdered ginger, verjuice [sour grape juice], salt and a little saffron. Serve.*

Harleian MS 4016, c. 1450

COMPOST. *Take parsley roots, parsnips, radishes; scrape them and wash them clean. Take turnips and cabbages, pared and cored. Take an earthenware pan with fresh water and set it on the fire; put all these in. When they have boiled, add pears and parboil well. Take all these things out and let cool on a clean cloth. Add salt; when it is cold, place it in a vessel; add vinegar, powder and saffron, and allow to sit for a night and a day. Take Greek wine and honey, clarified together; take Lombard mustard and whole currants, and powdered cinnamon, sugar and spice powder and whole anise seed, and fennel seed. Take all these things and place together in an earthenware pot, and take out and serve as required.*

The Forme of Cury, c. 1390

烤牛奶
Roasted milk

将甜牛奶倒入平底锅中，然后加入鸡蛋清并搅拌均匀，再撒上一些藏红花粉调色。加热至锅中牛奶变浓稠，倒出来用过滤网过滤掉多余的水分后挤压成型。冷却后就可以将它切成薄片放到烤盘上烤制了。

——《哈利父子文稿集》，1430

♣这道食谱创造出了一种清淡可口、需二次烹饪，还需在冷却后切片烤制的蛋奶糕。中世纪时，厨房中使用的还是加热炉和炭炉，火候的控制需要相当高超的技艺，所以说这道菜充分向我们展示了中世纪宫廷烹饪的复杂性。在没有控温的炉子和烤盘的情况下，厨师需要不间断地观察火势、添加燃料、移动烤盘和平底锅，使其远离或靠近火源等。

山楂花布丁
Hawthorn flower pudding

准备一些山楂花，煮熟并挤压后撕成小碎片，与杏仁乳混合，然后加入小麦淀粉、面包屑和黏米粉。再向其中加入足量的白糖或蜂蜜，将颜色调和成与山楂花颜色一样后即可食用。

——《哈利父子文稿集》，1430

♣山楂花在中世纪的食谱中非常常见，由山楂花制成的菜肴有一个共同的名字——“刺儿菜”，取这个名字大概是因为山楂树的刺棘。山楂花有股淡淡的杏仁味，在过去常被乡村的孩童们同山楂叶一起食用，并且被视为“简单的饭菜”。

ROASTED MILK

MILK ROASTS. *Take sweet milk and put it in a pan. Take eggs with all the whites and mix them and add them, colour it with saffron, and boil it so that it thickens; then strain it through a strainer, and take what remains and press it. When it is cold, cut it and slice it in thin slices and roast it on a griddle and serve.*

Harleian MS. 279, c. 1430

HAWTHORN FLOWER PUDDING

SPINEY. *Take the flowers of hawthorn; boil them and press them, grind them small, mix them with almond milk and add wheat starch, grated bread and rice flour. Add enough sugar or honey, and colour it the same as the flowers, and serve.*

Harleian MS. 279, c. 1430

蛇鸡兽

A mythical beast created from half a pig and half a cockerel

取一只阉割后的公鸡过沸水焯熟，然后将其内脏取出，拦腰斩断。同样地，取一头乳猪过沸水焯熟，然后将其内脏取出，拦腰斩断。用针线将阉鸡的前半身子与乳猪的后半身子缝合起来，然后将阉鸡的后半身子与乳猪的前半身子缝合起来，最后往它们的肚子里填充馅料，就像烤乳猪时在乳猪肚子里填料一样。填充完后插上烤叉开始烤制。烤好之后，涂上一层生蛋黄，撒上姜末和藏红花粉，再倒上欧芹汁，就可以在皇家盛宴上用它来招待宾客了。

——《哈利父子文稿集》，1430

♣神话里的蛇鸡兽有着鸡的头和鸡的爪子，还有爬虫、龙或蛇的尾巴。在英格兰，蛇鸡兽经常容易和蛇怪混淆。蛇怪是一种从鸡蛋里孵化出来的蛇，传说中看它一眼就会被它杀死。这个食谱仿佛创造出了一种转基因物种，将两种不同动物的身体缝在一起，做成了一种有点怪异却可以食用的专供皇室的菜式。

A MYTHICAL BEAST CREATED FROM HALF A PIG AND HALF A COCKEREL

COCKATRICE. *Take a capon [a castrated cockerel], and scald him; draw [eviscerate] him completely, and cut him in two around the waist. Take a pig, and scald him, and draw him in the same manner, and cut him also around the waist. Take needle and thread, and sew the fore part of the capon to the hind part of the pig, and the fore part of the pig to the hind part of the capon, and then stuff them as you stuff a pig. Put them on a spit and roast them, and when they are done enough, gild them with yolks of eggs, and powdered ginger and saffron, then pour over the juice of parsley; and then serve it for a royal feast.*

Harleian MS. 279, c. 1430

水罐猪肉

An artificial pitcher
made of pork, cheese and bread

取新鲜的瘦猪肉，煮熟后切成小块。然后准备一些藏红花粉、生姜、桂皮、食盐、高良姜（生姜的一种）、过熟干酪和面包屑，放入研钵中碾碎。把猪肉块放入刚刚碾碎的调味品中，让每一小块肉都包裹上调料，然后裹上一层生蛋液。取一个长形水罐，在内壁上涂上一层食用油，将调好味的猪肉块放到其中。将一块干净的布对折几次，盖住罐口，沿罐口边缘密封好，然后放入锅中隔水炖煮。煮好后打开水罐，取出罐中的猪肉块，插上烤串在火上烧烤。将藏红花粉、高良姜、桂皮和面粉混合在一起碾碎，再加入生鸡蛋调成糊状，然后加入足量的亚历山大白糖。当猪肉烤好后，涂上糊状调味料即可食用。

——《哈利父子文稿集》，1430

♣中世纪的烹饪书里全都是一些重口味的食谱，这些菜非常微妙，人们通常会用这些菜来戏弄宾客或者让宾客感到惊喜。这些菜形式各异，从糖人到栩栩如生的烤动物，再到形态不一的凉粉和果冻，变幻莫测。在英国，当时的大部分白糖都是通过亚历山大港和其他地中海东部的港口进口的，而糖的原产地在遥远的东方。

AN ARTIFICIAL PITCHER MADE OF PORK, CHEESE AND BREAD

APPRAYLERE. *Take the flesh of lean pork, and boil it well; and when it is boiled, cut it small; take saffron, ginger, cinnamon, salt, galangal [a kind of ginger], old cheese, grated bread, and grind it small in a mortar; then put the flesh into the spice mixture, and see that it is well ground, mix it up with raw egg. Then take a long pitcher, well rinsed out with grease, and fill the pitcher with the stuffing, and take a piece of clean canvas, and double it as much as you can over the mouth of the pitcher, and bind it fast about the rim, and put it to boil with your large joints, in a lead vessel or cauldron, so that it is well boiled. Then take your pitcher, and break it, and save your stuffing; and have a good spit, and spit it thoroughly, and lay it to the fire; and then have a good batter of enough spices, saffron, galangal, cinnamon and flour, and grind small in a mortar, and temper it with raw egg, and apply to it enough sugar of Alexandria; and as it dries, baste it with batter, and then serve.*

Harleian MS. 279, c. 1430

煮鲦鱼
Boiled minnows

首先，将鲦鱼或泥鳅清洗干净。用大量的麦芽酒和欧芹做成酱汁，当酱汁沸腾后，撇去上面的浮沫，浇到鱼身上，继续蒸煮。最后如果你想的话，可以加上一点藏红花粉和香草汁，然后趁热食用。

——《哈利父子文稿集》，1450

♣因鲦鱼的体形太小，英国人很少食用它，但在食物短缺的时候，鲦鱼因富含蛋白质而备受人们喜爱。是否能获取足够的粮食取决于天气状况和交通运输条件，当时的交通运输还未得到充分发展，运费十分昂贵，所以粮食经常短缺，特别是在冬天的那几个月。到了年底，很多饲养的动物都会被屠宰掉，因为在冬季已经没有足够的饲料来饲养它们。贫穷的人可能会靠捕捉河里或沿海的鱼过活，而富裕的人会依赖养殖在池塘中的鱼度日。有趣的是，流行于中世纪而后又逐渐“失宠”的那些菜在二十世纪两次世界大战粮食定量供应之际，又重新得到了人们的青睐。

BOILED MINNOWS

MINNOWS OR LOACHES BOILED. *Take minnows or loaches, and pick them clean; and make sauce of a good quantity of ale and parsley. And when it comes up to the boil, skim it clean, and cast in the fish and let it boil. And, if you wish, add a little saffron, and the sauce is a sauce verte [herb sauce]. And then you shall serve it hot.*

Harleian MS. 4016, c. 1450

海豚小麦粥

Porpoise with wheat porridge

取优质杏仁，洗净后碾碎，倒入水中，然后用过滤器过滤到装着牛奶的容器里。取小麦，将其碾碎，确保去除掉所有的小麦壳，然后将小麦放入清水中加热煮沸，直至麦粒完全煮熟，然后倒入准备好的杏仁牛奶，继续加热。当小麦粥变浓稠后，加入白糖、藏红花粉和食盐。准备一条鼠海豚，像处理鲑鱼那样切成大块，焯熟后切成薄片，与小麦粥一起食用。

——《哈利父子文稿集》，1450

鼠海豚是海豚家族里众多成员中的一员，它的肉在中世纪十分珍贵，只有在宫廷、贵族和教堂中位高权重的神职人员的餐桌上才会看到。虽然鼠海豚是一种哺乳动物，但却经常被当作鱼类进行烹煮，因此鼠海豚也可以在斋戒日食用。现在，对于足够虔诚的人来说，只有星期五才是吃鱼日，但在过去吃鱼日却有很多，包括星期三和星期六还有一年中其他大大小小的斋戒日。牛奶麦粥是一种煮沸的小麦粥，通常与鼠海豚肉一同食用。十八世纪以来，人们在食用各类烤肉时也喜欢搭配牛奶麦粥。如果有人想要创新这个菜谱的话，彼得·布里尔（Peter Brears）建议用金枪鱼代替鼠海豚。

PORPOISE WITH WHEAT PORRIDGE

FRUMENTY WITH PORPOISE. *Take good almonds, and wash them clean, and grind them in a mortar and draw them with water through a strainer into milk, and place it in a vessel. And then take wheat, and grind it in a mortar; so that all the hull is removed, and boil it in fresh water until it is well broken up and has boiled enough. Then take it from the fire and add the milk and let it boil. And when it is boiled enough and thick, add sugar; saffron and salt; then take a porpoise, and joint him as you would a salmon, and boil him in fresh water. And when it is done, cut it, and slice it into fair pieces, and serve with the frumenty, and pour hot water into the dish.*

Harleian MS. 4016, c. 1450

烤孔雀
How to roast a peacock and serve him in his skin

取一只孔雀，折断“他”的脖子，切断“他”的喉咙，将羽毛和皮一同剥掉，但要使孔雀的头部依然与整张孔雀皮连接在一起，不要破坏羽毛和孔雀皮的完整性。像处理鸡一样将孔雀的内脏取出，保证脖子骨头完整。将处理好的整只孔雀插到烤叉上烤制，脖子朝上，腿朝下，如同孔雀依然活着坐在那里。完全烤好后取下孔雀，自然冷却，再将整张孔雀皮绕在烤好的孔雀上，看起来像孔雀还活着一样，然后就可以端上餐桌享用了。要不然就不带孔雀皮，像吃烤鸡那样食用就好。

——《哈利父子文稿集》，1450

孔雀肉是一种在当时只有在辉煌盛宴上才能吃得到的肉类。孔雀并不是不列颠群岛土生土长的一种动物，所以必须通过饲养孔雀才能达到以下两个目的：一是用引人注目的孔雀来提高皇室的地位，二是用孔雀肉设宴款待尊贵的客人。几乎所有早期的食谱都有这种烹饪教程，教大家用带有羽毛的皮重新装饰烤好的孔雀，将它的尾巴展开，孔雀冠戴好，就像孔雀还是活的一样。孔雀肉据说有一种介于鸡肉和野鸡肉之间的味道，但是却又干又硬。食物学家彼得·布里尔曾提及，中世纪用于款待宾客的孔雀的尾巴都很短，一般是十八个月之内的幼孔雀，因此肉质柔软多汁。英国皇室从十六世纪末期开始不再食用孔雀肉了。

HOW TO ROAST A PEACOCK AND SERVE HIM IN HIS SKIN

PEACOCK ROASTED. *Take a peacock, break his neck, cut his throat, and flay the skin and the feathers together, with the head still on the skin of the neck, keeping whole the skin and the feathers together; draw [eviscerate] him like a chicken, keeping the neck bone whole. Roast him, setting the bone of the neck above the spit, as if he were sitting alive; and the body above the legs, as if he were sitting alive. When he is roasted enough, take him off, and let him cool; and then wind the skin with the feathers and the tail about the body, and serve him as if he were alive. Or else pluck him clean, and roast him, and serve as you do a chicken.*

Harleian MS. 4016, c. 1450

烤火车
Imitation entrails

取枣和无花果，将它们切成薄如硬币的片，再取大颗葡萄干和白杏仁，用针线将这四种水果一个接一个地穿起来，穿到人的身高那么长，然后将串好的水果串一圈圈绕到烤叉上。取一夸脱的红酒或麦芽酒，倒入适量面粉，搅拌成糊状，再加入适量生姜粉、白糖、藏红花粉、丁香粉和食盐。调味面糊稠度要适中，不能太稠也不能太稀，以可以从中间划开的稠度为最佳。将烤叉放于炭火上开始烤制，在翻烤时一遍遍地将调味面糊涂在上面，直至调味面糊将所有水果完全覆盖。当你涂调味面糊的时候，最好拿一个盘子在下面接住滴落的面糊。完全烤好后，它看起来就像动物的内脏。将其从烤叉上整个取下，横向切成食指的长度，趁热食用。

——《哈利父子文稿集》，1450

♣这道食谱是中世纪众多制作假冒烤肉的菜谱之一，目的是模仿出烤制的动物内脏的样子。在这道食谱中，调味面糊要完全裹住水果串，这样做出来的东西看起来才能和动物内脏非常地相似。食物历史学家伊万·戴（Ivan Day）曾成功地升级了这道菜的制作流程，他说："当厨房满是藏红花粉和烤制的调味面糊的味道，其表面呈现出金黄色的时候，就可以停止烤制了。如果线仍旧很牢固地绑在'烤火车'的两端，就小心地将线从中抽出来。"

IMITATION ENTRAILS

TRAIN ROAST. *Take dates and figs, and slice them as thin as a penny; and then take large raisins and blanched almonds, and prick them through with a needle onto a thread as long as a man, first one fruit and then another fruit; and then bind the thread with the fruit about a round spit, along the spit in the manner of a haslet. And then take a quart of wine or ale, and fine flour, and make a batter, and add powdered ginger, sugar, saffron, powdered cloves and salt; and make the batter neither runny nor thick, but in between, so that it will cleave. And then roast the train on the fire, on the spit; and then drop the batter on the train as it turns on the fire, until the fruit is hidden by the batter. As you drop on the batter, hold a vessel underneath, to catch any spilt batter, and when it is roasted well it will look like a haslet. And then take it up from the spit whole, and cut it in good pieces of a span's length, and serve a piece or two in a dish all hot.*

Harleian MS. 4016, c. 1450

秘制烤鸡

To make one capon into two by blowing up the skin

取一只阉鸡，将鸡皮小心地剥下来，保持其完整不破，洗净。剥下来的鸡身切成块，洗净，再准备一些猪肉块。将猪肉、鸡肉和优质香料混合在一起填充到剥开的鸡皮中，用针线缝合好后煮至半熟。然后给鸡身抹上食用油，再均匀地抹上用杏仁乳和小麦淀粉混合做成的面糊，放到火上烤制。烤制时往鸡身上涂抹藏红花粉，使其上色。烤熟后即可享用。

——《贵族烹饪法》，1468

月光鸡蛋

Eggs in moonshine

准备一碗玫瑰水和一碗糖水，将它们倒入锅中煮沸。取八个或九个蛋黄，一个一个下入锅中，成型后盛出，撒上一点肉桂粉和白糖即可食用。

——《一本烹饪新书》，1545

☙这道清淡可口的菜肴需要把火候控制得十分精准。几乎所有中世纪的菜肴都是通过明火烤制的，但是还有很多酱料和一些煮熟或煎炸的食物是在平底锅中制作的。平底锅一般放在放有木炭的陶器桶上进行加热，我们称之为火锅。厨师可以用火锅制作酱料、煎蛋卷或做月光鸡蛋，这些都需要控制好火候。现在我们可能仍会在一些复古的法国餐厅看到这种古老的火锅，这些法国餐厅坚持在饭桌上制作薄卷饼，不过他们的火锅一般不再使用木炭，而是通过燃气或电进行加热。

TO MAKE ONE CAPON INTO TWO BY BLOWING UP THE SKIN

TO MAKE CAPON IN CASSOLONT. *Take a capon and scald him and open the skin behind the head and blow the skin up with a quill and raise it all over. Then take pork and chicken flesh and good spices and make a stuffing, and sew the skin up and parboil it. Then roll the capon and lard it with fat and make a batter of almond milk and wheat starch and colour it with saffron at the fire and baste it and serve it.*

A Noble Boke off Cookry, 1468

EGGS IN MOONSHINE

TO MAKE EGGES IN MONESHYNE. *Take a dyche of rosewater and a dyshe full of suger, and set them upon a chaffyngdysh [chafing dish], and let them boyle, than take the yolkes of viii or ix egges newe layde and putte them therto everyone from other, and so lette them harden a lyttle, and so after this maner serve them forthe and cast a lyttle synamon [cinnamon] and sugar upon them.*

Proper New Booke of Cokerye, 1545

雪花碟
A dishful of snow

准备四品脱香甜浓厚的奶油、八个鸡蛋的蛋清、一匙玫瑰水以及一碗糖水。将蛋清搅拌均匀后倒入奶油、玫瑰水和糖水的混合物中。取一根干净的木棍，用木棍用力搅打后过滤，做出慕斯。取一个苹果，摆在盘子中间，在苹果上插入一根粗的迷迭香的枝条，然后将做好的慕斯倒在迷迭香和盘子上，就可以享用了。

——《一本烹饪新书》，1545

这道赏心悦目的食谱抓住了大自然的特点，开始先用蛋清、奶油、玫瑰水和糖水做成蛋白酥，然后通过搅打和过滤做出轻薄而香甜的慕斯。插在苹果上的迷迭香枝条摆放在盘子中央，被慕斯覆盖的枝条和盘子营造出一种被大雪覆盖的树木孤零零地立在茫茫雪地的景象。

A DISHFUL OF SNOW

TO MAKE A DYSCHEFULL OF SNOWE. *Take a pottell [four pints] of swete thycke creame and the whytes of eyghte egges, and beate them altogether wyth a spone, then putte them in youre creame and a saucerfull of Rosewater, and a dyshe full of Suger wyth all, then take a stycke and make it cleane, and than cutte it in the ende foure square, and therwith beate all the aforesayde thynges together, and ever as it ryseth take it of[f] and put it into a Collaunder, this done take one apple and set it in the myddes of it, and a thicke bushe of Rosemary, and set it in the myddes of the platter, then cast your Snowe uppon the Rosemarye and fyll your platter therwith. And yf you have wafers caste some in wyth all and thus serve them forthe.*

Proper New Booke of Cokerye, 1545

煮鞋子

To cook shoes

取牛臀肉，放到开水中煮一到两个小时，然后向锅中加入足量的卷心菜，与牛臀肉一起再煮三个小时。之后，再加入两只鸽子、水鸭或鹧鸪等禽类，一同煮熟后加入食盐即可享用。

——《一本烹饪新书》，1545

❧很难看出来这道菜中的鞋子在哪，做菜用的食材也没有一样看起来像脚上穿的鞋子，并且最后做成的菜也与鞋子毫不相干。考虑到十六世纪的单词发展到现在意义可能发生了改变，我们也只能猜测一下它的原始含义了。

TO COOK SHOES

TO MAKE SHOES. *Take a rumpe of beife and let it boyle an hower or two and put therto a gret quantitie of cole wurtes [cabbages] and let theim boile togither thre[e] howers then put to them a couple of stockdoves or teales [a type of duck], fesand, partrige or suche other wylde foules and let them boyle all togither then ceason them with salte and serue them forth.*

Proper New Booke of Cokerye, 1545

布丁兔

A rabbit with a pudding in his belly

取一整只兔子，剥皮，留住兔子耳朵并清洗干净。准备一些面包屑、切碎的板油和无核小葡萄干。再准备一些上好的香草、薄荷、欧芹、菠菜（或甜菜）、甜墨角兰，剁碎，然后用丁香、肉豆蔻粉、白糖、一丁点奶油、盐、蛋黄以及少量红枣碎调味。将所有食材和调料混合在一起，塞到兔子的肚子里，把肚子缝起来后放入锅中炖煮，煮一会后加入羊肉汤，或者一开始直接将兔子放到羊肉汤里一起煮。再加入醋栗或者葡萄干、淡黄油、酸葡萄汁、盐、面包屑和一点糖，煮好后就可以和面包或吐司片一起放到盘子里食用了。

——《一本烹饪书》，1591

水果布丁鱼

A carp with fruit pudding in his belly

将鲤鱼的鱼子煮熟后剁成鱼子酱。再将面包屑、小葡萄干、红枣碎、桂皮、白糖、丁香粒、肉豆蔻粉、胡椒粉和适量的盐搅拌在一起，做成水果布丁。把一小把鼠尾草放入水中慢炖，然后将三个或四个鸡蛋黄和一个鸡蛋清打成蛋液，倒进鼠尾草汤里调味。在剁好的鱼子酱中放入一点奶油和玫瑰花水。然后将做好的布丁放到鲤鱼的肚子里，放入鼠尾草水中炖煮，加一点盐。当鲤鱼快煮熟的时候，另外准备一个小锅，倒入一些鱼汤，以及一些鱼子酱、白葡萄酒、一片黄油、三到四个洋葱、足够的肉豆蔻粉、足够的胡椒粉、小葡萄干、三颗或四颗枣，一起炖煮。当大锅里的鱼煮入味后，放入足量的菠菜，适量蛋液，然后加入小锅中炖好的洋葱。如果味道太辣的话，加点白糖。最后将鲤鱼放到面包片或吐司片上，将煲好的汤倒在上面，就可以食用了。

——托马斯·道森，《好主妇是这样的》，1596

✤很多十六世纪和十七世纪的食谱都会在鱼或其他动物的肚子里塞上布丁，这种布丁在我们如今看来是一种填充馅料或者加料肉馅。这种馅料有很多用处：一是使外层食材的肉质变得更加香滑可口；二是利用放在食材肚子里的肉吸收多余的油脂。在当时对很多家庭来说，鱼和肉还是非常珍贵的，填充馅料的话会增加鱼和肉的用量，因此很多人在做的时候并不填充馅料。这类复杂的菜肴经常可以在好莱坞电影中看到。都铎王朝时期的食物经常是用一些别的肉和鸡肉混合做成的，在电影中，饥饿的国王和大臣们手嘴并用，将肉撕开后就将剩余的部分扔给身后的奴隶和趴在地上的狗了。

A RABBIT WITH A PUDDING IN HIS BELLY

TO BOYLE A CONY WITH A PUDDING IN HIS BELLY. *Take your cony [rabbit] and fley [skin] him, & leave on the eares and wash it faire, and take grated bread, sweete suet minced fine, corance [currants] and some fine hearbs, peneriall [pennyroyal]; winter savery; percely [parsley], spinage or beets, sweet margeram, and chop your hearbs fine, and season it with cloves, mace and sugar, a little creame and salt and yolks of eg[g]s, and dates minst fine. Then mingle all your stuf togither, and put it in your rabets belly and sowe it up with a thred, for the broth take mutton broth when it is boyled a little, and put it in a pot wheras your rabet may lye long waies in it, and let your broth boile or ever you put it in, then put in gooceberies or els grapes, corance and sweet butter, vergious [sour grape juice], salt, grated bread and sugar a little, and when it is boyled, lay it in a dish with sops [slices of bread or toast]. And so serve it in.*

A Book of Cookrye, 1591

A CARP WITH FRUIT PUDDING IN HIS BELLY

TO BOIL A CARP IN GREEN BROTH WITH A PUDDING IN HIS BELLY. *Take the spawn [roe] of a carp, and boil and crumble it as fine as you can. Then take grated bread, small raisins, dates minced, cinnamon, sugar, cloves, mace, pepper and a little salt, mingled together. Take a good handful of sage and boil it tender, and strain it with three or four yolks of eggs, and one white. Put to the spawn with a little cream and rose water. Then take the carp and put the pudding in the belly and seethe him in water and salt. When he is almost boiled take some of the spawn and of the best of the broth, and put it into a little pot with a little white wine, a good piece of butter, three or four onions, whole mace, whole pepper, small raisins, and three or four dates. When it is a good deal sodden, put in a good deal of seeded spinach, and strain it with three or four yolks of eggs, and the onions that you put in your broth. If it is too sharp put in a little sugar. And so lay your carp upon sops [slices of bread or toast] and pour your broth upon it.*

Thomas Dawson, The Good Housewife's Jewel, 1596

红烧麻雀吐司

Stewed sparrows on toast

取四品脱麦芽酒，或者自定义用量，放到火上煮沸。然后放入去毛洗净的麻雀，将汤中的浮沫撇掉，再加入足量的洋葱、欧芹、切成小段的迷迭香、胡椒粉、藏红花粉、丁香和肉豆蔻粉。然后像炸鱼或烤鱼那样做炸面包或者烤面包，将麻雀放到做好的面包上，倒上麻雀汤。如果需要的话还可以加一片黄油和一点酸葡萄汁。

——《一本烹饪书》，1591

☙一些考古发现的麻雀碗和鸟窝锅可以证明，在过去的英国，特别是伦敦，麻雀一定比我们想象中的还要频繁地被人食用。鸟窝锅是一种状如碗的陶制锅，可以挂在房子外面，就像现在的鸟箱一样，麻雀可以栖居其中。鸟类可以通过锅的顶部进出，但是在锅的后面同样有一个可以进出的洞，这个洞足够大，房子的主人可以将手伸进去然后将窝中的鸟蛋或者雏鸟拿走。考古学家曾提到，这种鸟窝锅在上层社会的家庭里很常见，这可能说明麻雀在当时是富裕者的美食，而非穷人食不果腹时的选择。

STEWED SPARROWS ON TOAST

TO STUE SPARROWES. *Take good ale a pottel [four pints], or after the quantities more or lesse by your discretion, and set it over the fier to boyle, and put in your sparrowes and scum the broth, then put therin onions, percely [parsley], time, rosemary chopped small, pepper and saffron, with cloves and mace, a fewe. And make sippets [bread, fried or toasted] as you doo for fish, and laye the sparrowes upon with the said broth, and in the seething [boiling or cooking] put in a peece of sweet butter, and vergious [sour grape juice], if need be.*

A Book of Cookrye, 1591

金币瓦罐鸡汤
A cockerel distilled with gold for treating consumption

取一只不太老的大红公鸡，杀死后剥皮，切成小块，然后将每块鸡肉都做脱骨处理。将处理好的鸡肉放到锅中，然后在其中按四比一的比例加入茴香根、欧芹、菊苣、紫罗兰叶和一定数量的玻璃苣，以及磨碎的香草、醋栗、肉豆蔻、大茴香子和切成片的甘草，进行调味。再将调过味的鸡肉全部装入小瓦罐里，加入四分之一的玫瑰水、一品脱白葡萄酒和两到三颗枣。再加入一枚金币（这样会更好）和半磅西梅干，盖上盖子，并用面团将瓦罐封住。将瓦罐放到锅中隔水炖煮十二个小时。十二个小时过后，取出并打开瓦罐，将鸡汤倒入锅中给体质虚弱的人早晚食用。

——《一本烹饪书》，1591

♣这道菜谱可以用香草、香料和无害的金币煮出公鸡体内的精华和营养。这种药用“精华”可以补足体质虚弱的人的身体亏空。《神圣的医书》（*The Book of Holy Medicines*）一书的作者亨利·格罗斯蒙特（Henry of Grosmont，1310—1361）将大红公鸡与基督联系在一起，写道：“大红公鸡就是你啊，最亲爱的耶稣，就像我之前说的，你就是救世主啊。”

A COCKEREL DISTILLED WITH GOLD FOR TREATING CONSUMPTION

TO STILL A COCK FOR A WEAKE BODY THAT IS CONSUMED. *Take a red cock that is not too olde, and beate him to death, and when he is dead, fley [skin] him and quarter him in small peeces, and bruse the bones everye one of them. Then take roots of fenell, persely [parsley], and succory [chicory], violet leaves, and a good quantitye of borage, put the cock in an earthen pipkin [cooking pot] and betweene everye quarter some rootes, hearbes, corance [currants], whole mace, anis seeds, being fine rubbed, and licorice being scraped and sliced, and so fill your pipkin with al the quarters of the Cocke, put in a quarter of a pinte of rosewater, a pinte of white wine, two or three dates. If you put in a peece of golde, it will be the better, and halfe a pound of prunes, and lay a cover upon it, and stop it with dough, and set the pipkin in a pot of seething [boiling] water, and so let it seethe twelve houres with a fire under the brasse pot that it standeth in, and the pot kept with licour twelve houres. When it hath sodden so many houres, then take out the pipkin, pul it open, and put the broth faire into a pot, give it unto the weak person morning and evening.*

A Book of Cookrye, 1591

酿萝卜
Fruit pudding in a hollowed-out turnip

取一根白萝卜，用温水洗净，去皮，然后像做裹馅胡萝卜那样将白萝卜挖成空心。将面包屑、苹果碎、醋栗、去壳的白煮蛋混合在一起，并加入白糖、桂皮、生姜、白煮蛋的蛋黄调味，做成馅料，最后将做好的馅料填充到空心萝卜里。锅中加水煮沸，放入加馅的萝卜，然后加入一块淡黄油、红葡萄酒、少许醋、迷迭香以及肉豆蔻、白糖、醋栗和红枣碎。当所有东西都煮软后，即可食用。

——《一本烹饪书》，1591

♣这种萝卜的制作方法十分特别，和酿苹果的做法非常相似。酿苹果作为一种甜品曾经非常受欢迎，但如今却很少能吃到。这道食谱中提到“像做裹馅胡萝卜那样将白萝卜做成空心”，对此，《一本烹饪书》中确实有裹馅胡萝卜这一菜谱，但是我们都铎王朝时期的祖先们在那时已经会为了裹馅而将胡萝卜做成空心的了？或者只是单纯地在做与胡萝卜有关的其他菜时将胡萝卜芯去掉？那个时期的胡萝卜的体形还不是很大，如今这么大的胡萝卜是在那之后才有的，植物选择和培育才使得我们如今的那些“大型”蔬菜得以出现。在那个时期的胡萝卜也不是我们现在经常看到的橘色，而是黄色、白色、红色甚至是紫色的。

FRUIT PUDDING IN A HOLLOWED-OUT TURNIP

HOW TO MAKE A PUDDING IN A TURNEP ROOT. *Take your turnep root, and wash it fair in warm water, and scrape it faire and make it hollow as you doo a carret roote, and make your stuffe [stuffing] of grated bread, and apples chopt fine, then take corance [currants], and hard egs, and season it with sugar, sinamon, and ginger, and yolks of hard egs and so temper [mix] your stuffe, and put it into the turnep, then take faire water, and set it on the fire, and let it boyle or ever you put in your turneps, then put in a good peece of sweet butter, and claret wine, and a little vinagre, and rosemarye, and whole mace, sugar, and corance, and dates quartered, and when they are boyled inough, then will they be tender, then serve it in.*

A Book of Cookrye, 1591

葡萄牙泡芙
Portuguese puffs

取一夸脱蜂蜜，加热至沸腾后撇掉浮沫，然后加入足量且过滤好的饼干粉以及丁香粉、生姜、肉桂粉、大茴香子和白糖，继续加热至你需要的黏稠状。因为用来做泡芙的面团要非常地成形，所以要和好的淡黄油与面粉混合在一起做成面团。做好的面团不能再随意揉捏。

——《一本烹饪书》，1591

❧这道食谱看起来像缺失了后面的一些制作步骤，因为后来的翻译本中清楚地说明了它是一种香甜的泡芙状的甜点。一些小的状似甜甜圈的泡芙可以被油炸，而且是空心的，很轻，在法国被称作空心油炸糕点，传入英格兰后被称作法式油炸泡芙（见本书第60页）。到了维多利亚时代，这种糕点又被称作“修女的悲叹”。至于葡萄牙的泡芙，则可能与西班牙状如甜甜圈的油条有一定的渊源。

PORTUGUESE PUFFS

TO MAKE FARTS OF PORTINGALE. *Take a quart of life hony, and set it upon the fire and when it seetheth [boils] scum it clean, and then put in a certaine of [sufficient] fine biskets well serced [sieved], and some pouder of cloves, some ginger, and powder of sinamon [cinnamon], annis seeds and some sugar, and let all these be well stirred upon the fire, til it be as thicke as you thinke needfull, and for the paste for them take flower as finelye dressed as may be, and a good peece of sweet butter, and woorke all these same well togither, and not knead it.*

A Book of Cookrye, 1591

海冬青蜜饯

Candied sea holly

将海冬青煮至绵软，去核，将其放进滤锅中滤净，待海冬青冷却。准备一些稀糖浆，加入海冬青茎叶的部分，冷藏放置三日，余下的海冬青的根部也一起冷藏。三日后，将糖浆（再加入一点新糖浆）加热。再过三日，重新加热，待冷却后，放入冷藏的海冬青的根部后保存。这样保存的海冬青根因为没有同糖浆一起加热过，所以口感更加软嫩。

——休·普拉特，《主妇的乐趣》，1602

♣众所周知，海冬青蜜饯同藏红花、牡蛎以及被人遗忘的埃塞克斯硬奶酪一样是埃塞克斯郡[①]曾经的主要食品。海冬青，即海滨刺芹，现在仍然沿着埃塞克斯以及其他城市的海岸生长，它的根部在过去经常被蜜制成“亲嘴蜜饯”，但是如今人们不再挖它的根部来食用了。从“亲嘴蜜饯”这个名字就可以看出蜜制的海冬青根具有催情的效果，莎士比亚对此也了如指掌，在莎士比亚的《温莎的风流娘儿们》（*The Merry Wives of Windsor*）中，福斯塔夫就要求天空应该“降下亲嘴蜜饯和雪冬青”。

---

① 埃塞克斯郡（Essex）是英国东英格兰大区的郡，位于英国东南部。

CANDIED SEA HOLLY

HOW TO PRESERVE ERINGO ROOTS, AENULA CAMPANA, AND SO OF OTHERS IN THE SAME MANNER. *Seethe [boil] them till they be tender: then take away the piths of them, and leave them in a colander till they have dropped as much as they will: then having a thin sirup ready, put them being cold into the sirup beeing also cold, and let them stand so three daies, then boyle the sirup (adding some fresh sirup to it; to supply that which the rootes have drunke up) a little higher: and at three daies end, boyle the sirup againe without any new addition, unto the full height of a preserving sirup, and put it in your rootes, and so keep them. Rootes preserved in this manner, will eate very tender, because they never boyled in the sirup.*

Hugh Plat, Delightes for Ladies, c. 1602

糖皮兔子

Rabbits, woodcock and other animals cast in sugar

首先将鱼鳔做成的白明胶溶解于清水中，溶解后可以再加入一点玫瑰水。取漂白的杏仁，将其碾成做杏仁饼用的碎杏仁大小，然后用奶油（牛奶也可以，但奶油做出来的会更脆）和玫瑰水调和，再加入一些白糖，倒入一些温水，最后将溶解的白明胶放入其中（记住：白明胶放得越多，最后做出来的食物就会越硬），这样糖皮就做好了（小提示：一夸脱奶油，一夸脱杏仁，两盎司鱼鳔，四或六盎司白糖对做糖皮来说比例最为适宜）。接着拿出准备好的兔子、丘鹬或其他形状的模具（如果模具是木头材质的话，要先给模具涂上甜杏仁油，如果是石头的或石膏的话就要涂上猪油），倒入做好的糖皮，按压成形。你还可以在上面撒上面包屑、桂皮和白糖，这样看起来就会像烤制或烘焙出来的。你也可以用擀面杖把糖皮擀成你想要的厚度。糖皮放不了太长时间，所以在做完之后的几天内就得吃掉。

——休·普拉特，《主妇的乐趣》，1602

♣这种用糖和模具做成的人或动物状的糖制品在中世纪和现代社会早期广受一些人的喜爱，这些人通常能够买得起进口的昂贵白糖。就像休·普拉特（Hugh Plat）所说的那样，做糖制品的模具可能是由木头或石膏做成的，而且一般能够被拆成两到三部分，将这几部分拼在一起就可以做成最后的糖制品。石膏模具通常由真实的水果和蔬菜以及其他东西铸成，甚至还可以是死去的动物和鸟类。最后做成的糖制品可能会被画上图案，用来装饰餐桌或在宴会上同最后一道菜一同端出。普拉特的食谱和现代的杏仁软糖的做法非常接近，而被做成水果、蔬菜、动物形状的杏仁软糖在欧洲还非常受欢迎。

RABBITS, WOODCOCK AND OTHER ANIMALS CAST IN SUGAR

A MOST DELICATE AND STIFF SUGAR PASTE, WHEREOF TO CAST RABBETS, PIGEONS, OR ANY OTHER LITTLE BIRDE OR BEAST, EITHER FROM THE LIFE OR CARVED MOULDS. *First dissolve issinglasse [a gelatine made from fish bladders] in faire water, or with some rose-water in the later end; then beat blanched almonds, as you would for marchpane stuff [marzipan], and draw the same with creame and rose water (milke will serve, but creame is more delicate): then put therein some powdered sugar; into which you may dissolve your issinglasse, being first made into gelly, in fair warm water (note, the more isinglasse you put therein, the stiffer your work will prove): then having your rabbets, woodcock, &c molded either in plaster from life, or else carved in wood (first anointing your wooden molds with oile of sweet almonds, and your plaister or stone moulds with barrows grease [fat from a castrated male pig]), pour your sugarpaste thereon. A quart of creame, a quarterne [quarter] of almonds, two ounces of isinglasse, and foure or six ounces of sugar, is a reasonable good proportion for this stuffe. You may dredge over your foule with crums of bread, cinamon and sugar boiled together, and so they will seem as if they were rosted and breaded. Leach [set foods] and gelly may be cast in this manner. This paste you may also drive with a fine rowling pin, as smooth and as thin as you please: it lasteth not long, and therefore it must bee eaten within a few daies after the making thereof. By this meanes, a banquet may bee presented in the forme of a supper, being a very rare and strange device.*

Hugh Plat, Delightes for Ladies, c. 1602

解渴软果糕
Thirst-quenching fruit pastilles for ladies

取半磅达马斯卡西梅干和一夸脱枣，将它们同一个蒸烤好的梨或者一些煮熟的果泥一起放到研钵里碾碎，然后将碾碎的东西按压到模具中，脱模后放入炉子中烤制，风干后再撒一些生姜粉，就可以在宴会上供客人享用了。

——休·普拉特，《主妇的乐趣》，1602

♣十七世纪早期的旅行者们担心途中河流、小溪或村井里的水有可能受到了污染，因此他们更愿意在到达旅馆后吃上一块软果糕来缓解口渴。那时人们更喜欢喝啤酒或者麦芽酒，因为经过蒸馏和发酵等步骤制作出来的酒会比水更加安全。

THIRST-QUENCHING FRUIT PASTILLES FOR LADIES

TO MAKE A PASTE TO KEEP YOU MOIST, IF YOU LIST NOT TO DRINK OFT; WHICH LADIES USE TO CARRY WITH THEM WHEN THEY RIDE ABROAD. *Take halfe a pound of Damaske prunes, and a quartern [quarter] of dates: stone them both, and beat them in a mortar with one warden [cooking pear] being rosted, or else a slice of old marmalade [cooked fruit paste]: and so print it in your moulds and dry it after you have drawne bread [from the oven]: put Ginger into it, and you may serve it at a banquet.*

Hugh Plat, Delightes for Ladies, c. 1602

蜜饯豆蔻

Nutmegs candied for three weeks then smashed out of their pots

准备一磅白糖、八勺玫瑰水、六便士重的阿拉伯树胶，将它们混合在一起，蒸煮至适宜浓度。然后将煮好的糖浆倒入罐子里，放入豆蔻、生姜等，拌匀。用黏土将罐口密封，防止空气进入，将罐子在高温下放置三周。三周后你得用锤子将罐子敲碎，否则很难取出里面的蜜饯豆蔻。当然，你也可以把豆蔻换成你喜欢的橘子或柠檬一类的东西。

——休·普拉特，《主妇的乐趣》，1602

在有冷冻类或罐头类的食品之前，用糖蜜制是能够将食物保存一年之久的主要方法之一。风干的生姜和豆蔻以及蜜制的生姜在很久以前就被引进到了英格兰。休·普拉特从国外找到了蜜饯豆蔻的做法，但是却很难做成功，因为硬如石头的豆蔻不可能在三周之内就有松软可口的口感。难道是因为休·普拉特得到的食谱只是断章取义的，或者他将成熟后风干的种子与肉豆蔻外层的果实混为一谈了？在东印度，一般会用糖腌制肉豆蔻外层的果实。当然，现在休·普拉特肯定无法获得十七世纪早期英格兰的新鲜肉豆蔻果实了，但在那时，即便是交通便利的情况下，运送这些东西耗时也不会短吧？

NUTMEGS CANDIED FOR THREE WEEKS THEN SMASHED OUT OF THEIR POTS

TO CANDIE NUTMEGS OR GINGER WITH AN HARD ROCK CANDY. *Take one pound of fine sugar, and eight spoonfuls of rose-water, and the weight of six pence of gum Arabique, that is cleere: boyle them together to such an height, as that, dropping some thereof out of a spoon, the sirup doe rope and runne into the smallnesse of an haire: then put it into an earthen pipkin [cooking pot]; wherein place your nutmegs, ginger, or such like: then stop it close with a sawcer, and lute [seal] it well with clay, that not aire may enter: then keepe it in a hot place three weeks, and it will candy hard. You must breake your pot with a hammer, for otherwise you cannot get out your candy. You may also candy orenges or lemmons in like sort, if you please.*

Hugh Plat, Delightes for Ladies, c. 1602

用树脂保存水果

Fruit preserved in pitch

露莓和黑莓有点相像。将其浸泡在热树脂中，在树脂差不多冷却但还未凝结和变硬之时，提着露莓的梗将它们从树脂中取出。这样处理过的露莓可以保存一整年。

——休·普拉特，《主妇的乐趣》，1602

♣在十七世纪初期，焦油和黏稠度更高的树脂是瑞典的主要出口商品。它们都是通过加热或者干馏松树做出来的，在制作的过程中会产生木炭和松脂等副产品。露莓是茄属植物中有剧毒的水果之一，也被称为颠茄，甚至还被叫作“坏男人的恶果”。显而易见，这个食谱是用来制作药物的。休·普拉特用树脂保存露莓的想法可能来自他对罗马作家卢修斯·科卢梅拉（Lucius Columella）的研究，卢修斯·科卢梅拉曾建议过用树脂来保存葡萄。普拉特宽泛地写了很多储存食物的方法，包括将龙虾包裹在海水浸泡过的布中，然后将其埋在沙子里；还有在石榴外面涂上蜡，然后挂起来。后者与现代零售行业给柠檬上蜡的操作相似，都可以延长水果的保鲜时间。普拉特甚至还建议将牛肉放到有破洞的桶中，然后将桶悬挂在海途漫漫的轮船尾部，以此来保存牛肉。

FRUIT PRESERVED IN PITCH

FRUIT PRESERVED IN PITCH. *Dwayberries that doe somewhat resemble blacke cherries, called in Latine by the name Solanum laethale, being dipped in molten pitch, being almost cold, and before it congeale and harden againe, and so hung up by their stalkes, will last a whole yeare.*

Hugh Plat, Delightes for Ladies, c. 1602

麦芽黄瓜

A cucumber stuffed with barley to attract flies

意大利人制作这个食谱所发挥的实用性和稀有性着实让我耳目一新。简单来说，就是在一根空心黄瓜里装满发芽的大麦粒。但首先要用木头或骨头做成的刺针给黄瓜戳上小洞，装上大麦粒之后，这些大麦粒就会填充到各个小洞里，这样就没有人能看出这是什么东西了。夏季，将制作完的麦芽黄瓜挂到房间的中央，就会吸引很多蝇虫落在上面，这样蝇虫就不会落到墙上挂的各种装饰或物件上了。

——休·普拉特，《主妇的乐趣》，1602

☙ 与其说这是一道食谱，不如说它是一个生活妙招。而休·普拉特，与其说他是食谱创作者，不如称其为食谱收藏者，更为贴切。他并没有详细说明装到黄瓜里的大麦如何能够快速发芽，从而掩盖黄瓜本身的模样。这样做出来的东西到最后肯定看起来非常像某个奇异的海洋生物，如同吊灯一样挂在房间的中央。毫无疑问，这个东西会随着黄瓜的变干和腐烂变得越来越奇怪，大麦芽最终也会枯萎凋谢。

A CUCUMBER STUFFED WITH BARLEY TO ATTRACT FLIES

To Keep Flies off your Paintings and Hangings. *An Italian conceipte both for the rareness and use thereof doth please me above all other: viz: pricke a cowcumber full of barley corns with the small spring ends outwards, make little holes in the cowcumber first with a wooden or bone bodkin, and after put in the grain, these being thicke placed will in time cover all the cowcumber so as no man can discerne what strange plant the same should be. Such cowcumbers to be hung up in the middest of summer rooms to drawe all the flies unto them, which otherwise would flie upon the pictures or hangings.*

Hugh Plat, Delightes for Ladies, c. 1602

烤黄油
Butter roasted on a spit

要将一磅的黄油烤得美味可口，首先要准备一磅淡黄油，与白糖、鸡蛋黄混合后打稠，然后拍打成形，放到烤架上用温火烤制。烤制的过程中要不断地往黄油块上撒面粉，防止黄油融化滴落。将黄油块烤至金黄色就可以取下装盘了，盘子里撒上一层白糖效果会更好。

——杰维斯·马卡姆，《英国主妇》，1615

♣在做这道菜时，马卡姆会加入优质面包屑、醋栗、白糖和食盐等食材，他还用同样的方法做过烤乳猪。烤制时，需要隔一段时间就在黄油块上撒一层面粉，这样黄油块外面就会形成一层甜甜的脆皮。食物学家伊万·戴（Ivan Day）曾参照这道食谱成功做出了烤黄油，据他说按这道食谱做出来的烤黄油非常美味。

BUTTER ROASTED ON A SPIT

TO ROAST A POUND OF BUTTER WELL. *To roast a pound of butter curiously and wel, you shal take a pound of swete butter and beat it stiffe with sugar, and the yolkes of eggs; then clap it roundwise about a spit, and lay it before a soft fire, and presently dredge it with the dredging before appointed for the pig; then as it armeth or melteth, so apply it with dredging till the butter be overcomed and no more will melt to fall from it, then roast it brown, and so draw it, and serve it out, the dish being as neatly trimmed with sugar as may be.*

Gervase Markham, The English Hus-wife, 1615

旅行携带的即食干醋
Dried instant vinegar for travellers

要制作能够随身携带的干醋，首先要准备一些小麦或者黑麦的叶片，放到研钵中，倒入你能找到的最高浓度的醋，将叶片碾碎至糊状，然后团成小球。将团成的小球放到太阳下晒干，当你想食用的时候就可以把晒干的小球切成片，放在酒中溶解，这样就可以获得高浓度的食醋了。

——杰维斯·马卡姆，《英国主妇》，1615

♣食醋中最重要的成分就是醋酸。马卡姆的做法能够将醋酸浓缩并脱水，这样做出的叶片泡在酒里不是真的将酒变成了醋，而是使酒变得像醋一样酸。在那个时期，食醋作为一种调味汁一般用来倒在热蔬菜盘中。这种做法延续至今，但现在人们一般只在吃薯片的时候撒上盐和醋。当然，如今我们也会用食醋调制沙拉。

DRIED INSTANT VINEGAR FOR TRAVELLERS

TO MAKE DRY VINEGAR. *To make dry vinegar which you may carry in your pocket, you shall take the blades of green corn either Wheat or Rye, and beat it in a mortar with the strongest vinegar you can get, till it come to a paste, then roule it into little balls, and dry it in the Sun till it be very hard, then when you have occasion to use it cut a little piece thereof and dissolve it in Wine, and it will make a strong vinegar.*

Gervase Markham, The English Hus-wife, 1615

鲦鱼艾菊蛋卷

Minnow, tansey, cowslip and primrose omelette

在春天，人们经常做好吃的鲦鱼艾菊蛋卷。将鲦鱼放入盐水中洗净，去掉鱼头和鱼尾，清除鱼内脏，然后与蛋黄、樱草花、迎春花以及少量艾菊一同煎炸，这样做出来的蛋卷比较好吃。最后可以与肉一起放入盘中，组成一道美味可口的菜肴。

——伊萨克·沃尔顿，《钓客清话》，1653

♣食谱里提到的艾菊是长得像布丁一样的花，这个名字源于菊蒿一类的香草。这种香草开有黄色的、纽扣大小的花，味道十分浓烈，所以在做菜的时候只需要放少量就足够了。因为艾菊很常见又讨人喜爱，所以早期用艾菊做成的菜谱有很多，甚至有些食谱里根本没有用上艾菊，但是食谱名里还是会出现艾菊。1974年，烹饪作家伊丽莎白·艾尔顿（Elizabeth Ayrton）在描写上述沃尔顿的这道食谱的时候提到，如果你在四月份的小溪中捕捉到了鲦鱼，然后在穿过一片花海回家的路上采摘了樱草和迎春花，你就会吃到英国春天的味道。

MINNOW, TANSEY, COWSLIP AND PRIMROSE OMELETTE

MINNOW TANSEY. *In the spring they make of them excellent minnow-tansies; for being well washed in salt, and their heads and tails cut off, and their guts taken out, and not washed after, they prove excellent for that use; that is, being fried with yolks of eggs, the flowers of cowslips, and of primroses, and a little tansy; thus used they make a dainty dish of meat.*

Isaak Walton, The Compleat Angler, 1653

芦笋炒乌龟
Sautéed tortoise with asparagus

将乌龟洗净，切成块，用少量黄油、欧芹和葱煎炒一下，调味，再用文火慢炖。注意在切乌龟的时候不要把它的苦胆切碎，并确保乌龟清洗干净。加热面包，与乌龟肉和炖出的汤水一同装到盘子中，盘子周围放上脆芦笋、蘑菇、松露、柠檬片，再淋上一些蘑菇汁，就可以享用了。

——拉瓦朗，《法国烹饪》，1653

♣由法国烹饪学家拉瓦朗所写的烹饪书在十七世纪中期被翻译成英文后声名大噪。我们可以肯定的是英国人对法国食物的痴迷始于拉瓦朗的这几本翻译而来的烹饪书。在十九世纪，不论是有钱人还是饭馆餐厅，都喜欢雇佣法国厨师或者制作法语菜单。关于上述这道特别的菜谱，拉瓦朗当时可以在法国获取乌龟这一食材，因为在法国本土乌龟很常见，但貌似英国厨师却很少有机会做这道菜。

SAUTÉED TORTOISE WITH ASPARAGUS

POTTAGE OF TORTOISE. *Take your tortoises, clean them and cut them into pieces, fry them with a little butter, parsley and spring onion having strained and seasoned them well, simmer them in a dish with some stock on the stove. Make sure you get rid of the turtle gall when you cut them up, to insure your tortoises are clean and cook well in the well-seasoned water. Simmer your bread also, and then garnish it with your tortoises and their sauce, with snapped asparagus around the dish, mushrooms, truffles, slices of lemon and mushroom juice and serve.*

La Varenne, The French Cook, 1653

法式油炸泡芙
Whore's Farts

用来做油炸泡芙的面糊要比一般面糊更加浓稠，因此需要加入更多的面粉和鸡蛋，用力搅拌均匀后将它分成小块。炸好后就可与白糖和开胃菜一起食用了。

——拉瓦朗，《法国烹饪》，1653

♣泡芙之所以叫泡芙，可能不是因为它们如空气般轻盈，而是因为制作此类食谱时都会用一个大的、像注射器一样的东西将面糊挤到油锅中，被挤出的那一刻会发出一种奇怪的声音。

烤水獭
Barbecued otter

在烤制之前给水獭涂上调料，烤完后如果味道不够重的话，就配上你想吃的酱料一起吃。因为大块的水獭肉不好入味，所以可以将其切成薄片或者在上面切上几刀帮助入味。还可以将烤好的水獭肉浸到调制好的酱料中炖煮，完全入味后就可以和配菜一同食用了。

——拉瓦朗，《法国烹饪》，1653

♣虽然当时乡村的人们经常会捕捉水獭，但是除了这道烤水獭的食谱外，很难再在英国的烹饪书中找到其他与水獭有关的食谱了。有趣的是，从佛兰德画家法兰斯·斯奈德（Frans Snyders）在十七世纪早期所作的画中可以看到人们在那个时期食用水獭的图景。画中描述的是一个鱼贩在摊位前摆满了各种鱼类、龙虾、螃蟹和贝类，其中在一只活力满满的海豹旁边还放着一只死气沉沉的水獭。

WHORE'S FARTS

A VARIATION ON PORTUGUESE PUFFS. *Make up your fritter batter stiffer than ordinary by increasing the flour and eggs, then set them out very small. When the fritters are cooked, serve them hot with sugar and a savoury water.*

La Varenne, The French Cook, 1653

BARBECUED OTTER

SEA-OTTER ON THE GRILL. *Dress the sea-otter and roast it. When it is done, make whatever sauce you like for it, provided it tastes strong and, because the large pieces don't readily take on a flavouring, split it or slice it on top. Simmer it in its sauce until it has soaked up almost all of it. Then serve it, garnished with whatever you have on hand.*

La Varenne, The French Cook, 1653

派对战火餐

Pies containing live frogs and birds, a pastry stag that bleeds, exploding ships and castles

用硬纸板做成船的形状，挂上旗帜和彩带并装上其他零碎物品，如枪支，用线固定。给做好的模型涂上一层糨糊，然后将它放在显眼的地方，配上作战需要的枪支弹药和战马，粘上装有甜水的鸡蛋壳，并在周围撒上食盐。

再用面团做一个雄鹿模型，鹿身上插着用硬纸板做成的箭，还要洒满波尔多葡萄酒。再用纸板做一座有防卫墙、铁闸门、大门和吊桥的城堡，也分别涂抹上糨糊。将两个模型按合适的距离摆放，好似正在军火中交战。

在每一个模型旁边都放上一个用面团做成的馅饼，馅饼里裹着活青蛙还有活鸟。做法是首先在面团里放入麦麸、藏红花粉和蛋黄，烘焙后在馅饼的下面开一个小洞，取出里面的麦麸，塞进活青蛙和活鸟，然后将小洞用原来的面团合上。将做好的所有东西端上餐桌，客人可以将雄鹿身上的箭拔出，雄鹿身上的葡萄酒就会像鲜血一样流出。

旁边的人会对此瞠目结舌，稍等片刻后，让客人将装有甜水的鸡蛋壳拿起，互相抛掷，此时战局就会显得紧张万分。这时你就可以提出大家或许想看下馅饼里裹着什么馅，当青蛙跳出来的时候，客人们就会惊跳而起，惊喜异常；这时再把另一个馅饼打开，让小鸟飞出，小鸟会本能地朝着火光飞去，把蜡烛扑灭。鸟飞蛙跳，非常热闹，必定会让现场的气氛更加活跃。

——罗伯特·梅，《成就厨师》，1660

PIES CONTAINING LIVE FROGS AND BIRDS, A PASTRY STAG THAT BLEEDS, EXPLODING SHIPS AND CASTLES

Triumphs and Trophies in Cookery, to be used at Festival Times, as Twelfth-day, & c.

Make the likeness of a ship in paste-board [cardboard], with flags and streamers, the guns belonging to it of kickses [odds and ends], bind them about with packthread, and cover them with close paste proportionable to the fashion of a cannon with carriages, lay them in places convenient as you see them in ships of war, with such holes and trains of [gun]powder that they may all take fire; place your ship firm in the great charger; then make a salt round about it, and stick therein egg-shells full of sweet water. Then in another charger have the proportion [model] of a stag made of course paste [pastry], with a broad arrow in the side of him, and his body filled up with claret-wine; in another charger at the end of the stag have the proportion of a castle with battlements, portcullices, gates and draw-bridges made of paste-board, the guns and kickses, and covered with course paste as the former; place it at a distance from the ship to fire at each other. At each side of the charger wherein is the stag, place a pye made of course paste, in one of which let there be some live frogs, in each other some live birds; make these pyes of course paste filled with bran, and yellowed over with saffron or the yolks of eggs, guild them over in spots… being baked, and make a hole in the bottom of your pyes, take out the bran, put in your frogs, and birds, and close up the holes with the same course paste… being all placed in order upon the table, before you fire the trains of powder, order it so that some of the ladies may be perswaded to pluck the arrow out of the stag, then will the claret-wine follow, as blood that runneth out of a wound. This being done with admiration to the beholders, after some short pause, fire the train of the castle, that the pieces all of one side may go off, then fire the trains, of one side of the ship as in a battel; next turn the chargers; and by degrees fire the trains of each other side as before. This done to sweeten the stink of powder, let the ladies take the egg-shells full of sweet waters and throw them at each other.

All dangers being seemingly over, by this time you may suppose they will desire to see what is in the pyes; where lifting first the lid off one pye, out skip some frogs, which make the ladies to skip and shreek; next after the other pye, whence come out the birds, who by a natural instinct flying in the light, will put out the candles; so that what with the flying birds and skipping frogs, the one above, the other beneath, will cause much delight and pleasure to the whole company.

Robert May, The Accomplish't Cook, 1660

A MODERN BALL BUFFET.

肉杂碎吐司

Lips, noses, udders, ox-eyes and sparrows on toast

将牛颚、牛唇、牛鼻、牛眼、麻雀切成一先令的大小，放到蒸锅中，再加入一些黄瓜丁、生牛乳、洋蓟、土豆、方形甜瓜块、肉豆蔻、两到三朵完整的丁香、火腿、小牛胰脏、一些百灵鸟或像麻雀一样的鸟类、牛眼、食盐、黄油、浓汤、嫩葫芦、白葡萄酒、葡萄、伏牛花果、醋栗、白煮蛋的蛋黄，一起用小火慢炖，炖的时候可以多加点蛋黄和酸葡萄汁来增加汤的浓稠度。炖熟后夹到法式吐司中，配上几片柠檬即可食用。

——罗伯特·梅，《成就厨师》，1660

❧罗伯特·梅的食谱记录的是十七世纪英格兰富裕家庭食用的菜肴。从食谱中我们可以清楚地了解到当时的富裕家庭并不排斥食用这些比较便宜的肉杂碎。如今，我们会把这些肉杂碎做成火腿或者加工成其他食品，而且很少有人会吃牛颚、牛唇和牛鼻。通过这道食谱可以得知，通过充分的准备和烹饪，这些动物肉杂碎也可以变成美味的菜肴。值得注意的是，这道食谱中还加入了土豆，土豆在早期食谱里很少被提及，因为直到十八世纪人们才开始习惯食用土豆这种蔬菜。

LIPS, NOSES, UDDERS, OX-EYES AND SPARROWS ON TOAST

TO STEW PALLETS, LIPS, AND NOSES OF BEEF. *Take them being tender boild and blanched, put them into a pipkin [cooking pot], and cut to the bigness of a shilling, put to them some small cucumbers pickled, raw calves udders, some artichocks, potatoes boil'd or muskmellon in square pieces, large mace, two or three whole cloves, some small links or sausages, sweetbreads of veal, some larks, or other small birds, as sparrows, or ox-eyes, salt, butter, strong broth, marrow, white-wine, grapes, barberries [fruit of the shrub berberis], or gooseberries, yolks of hard eggs, and stew them all together, serve them on toasts of fine French bread, and slic't lemon; sometimes thicken the broth with yolks of strained eggs and verjuyce [verjuice, sour grape juice].*

Robert May, The Accomplish't Cook, 1660

鱼肉馅饼

Fishy mince pies

取一条鲤鱼，清洗后去掉鱼骨和鱼刺。再取一条肥大的鳝鱼，做同样的处理。把它们剁碎后，加入胡椒粉、肉豆蔻粉、桂皮、姜和食盐，搅拌成馅，然后加入醋栗、香菜籽、切碎的橘子皮、六个或七个白煮蛋的鸡蛋黄、枣片和白糖，搅拌均匀。将调好的馅装进馅饼，然后在馅饼的底部涂上黄油，烘焙后撒上一层糖霜即可食用。

——罗伯特·梅，《成就厨师》，1660

在十七世纪上半叶，英国人仍旧很喜欢甜咸搭配的食物，甚至会在其中加入水果干，比如上述食谱中提到的枣片、醋栗和鱼肉的搭配。这道食谱与俄罗斯的长形大烤饼有点类似，这种烤饼也是将鱼肉与煮熟的鸡蛋结合在一起，有时候还会加点水果干。纵观十六世纪和十七世纪，包括肉饼和鱼肉馅饼在内的各类馅饼最后都会撒一层糖霜。馅饼在撒上糖粉后会重新回炉，这样表面就会形成一层亮晶晶的糖霜。

FISHY MINCE PIES

TO MAKE MINCED PIES OF CARPS AND EELS. *Take a carp being cleansed, bone it, and also a good fat eel, mince them together, and season them with pepper, nutmeg, cinamon, ginger, and salt, put to them some currants, caraway-seed, minced orange-peel, and the yolks of six or seven hard eggs minced also, sliced dates, and sugar; then lay some butter in the bottom of the pyes, and fill them, close them up, bake them, and ice them.*

Robert May, The Accomplish't Cook, 1660

治瘟疫的药酒
A cure for the plague

取芸香、龙牙草、艾草、白屈菜、鼠尾草、香脂草、艾蒿、海绿、金盏花、小白菊、地榆、酸模、切碎的土木香根、山萝卜、马先蒿、金母菊、薄荷、直立委陵菜、飞廉以及迷迭香，如果可以的话还可以加点当归。所有这些食材的用量都差不多，但迷迭香的用量要是其他食材的两倍。将所有食材切片，混合在一起，然后用优质的白葡萄酒浸泡三天三夜，白葡萄酒以刚好浸没食材为佳，浸泡的时候每隔半天或一天搅拌一次。浸泡完后就可以喝了，第一次不要喝太多，感觉有所好转后第二次就可以多喝一点。

——《博学的卡奈米·迪格比先生的密室大公开》，1669

☙1665年，传染病盛行，对当时的人来说，瘟疫是人们的心头大患。但是卡奈米·迪格比（Kenelme Digby）以及与他同时代的那些烹饪家还不知道，1665年爆发的大瘟疫实际上是英国最后一次大规模瘟疫了。

A CURE FOR THE PLAGUE

TO MAKE PLAGUE WATER. *Take rue, agrimony, wormwood, celandine, sage, balm, mugwort, dragons, pimpernel, marigold, fetherfew, burnet, sorrel, and elecampane-roots scraped and sliced small, scabious, woodbetony, brown may-weed, mints, avence, tormentil, carduus benedictus, and rosemary as much as any thing else, and angelica if you will. You must have like weight of all them, except rosemary aforesaid, which you must have twice as much of as of any of the rest; then mingle them all together and shred them very small; then steep them in the best white-wine you can get three days and three nights, stirring them once or twice a day, putting no more wine than will cover the herbs well; then still it in a common still, and take not too much of the first water, and but a little of the second, according as you feel the strength, else it will be sour. There must be but half as much elecampane as of the rest.*

The Closet of the Eminently Learned Sir Kenelme Digby Kt. Opened, 1669

旋塞啤酒

To bottle cock ale

将一只处理干净的公鸡放入锅中，然后加入八加仑麦芽啤酒，煮沸。再取四磅晒好的葡萄干、两个或三个肉豆蔻、三个或四个肉豆蔻干皮、半磅枣，用研钵碾碎，放入两夸脱匀兑而成的优质白葡萄酒中。将麦芽啤酒中的公鸡取出，与白葡萄酒混合，密封放置六到七天后装瓶，一个月后即可饮用。

——《博学的卡奈米·迪格比先生的密室大公开》，1669

♣塞缪尔·佩皮斯（Samuel Pepys）在他的日记中曾记载他在1663年2月2日喝了一杯旋塞啤酒，但他没有对这种酒做任何评价。因此，我们现在并不清楚公鸡在这种啤酒中所发挥的作用，也不知道如果换成普通的麦芽啤酒的话对那些喜欢尝试辣咖喱和香辣肉酱但又觉得这些还不够辣的“真男人”们来说会有什么不同。与此相类似，一些食谱中也会在装有麦芽酒的帆布袋中放入碾碎的鸡肉、鸡骨头等。然而，现代版的啤酒制作方法提醒道：“想要自己制作旋塞啤酒的话，需要防范啤酒中的微生物污染。”

TO BOTTLE COCK ALE

TO MAKE COCK-ALE. *Take eight gallons of ale, take a cock and boil him well; then take four pounds of raisins of the sun well stoned, two or three nutmegs, three or four flakes of mace, half a pound of dates; beat these all in a mortar, and put to them two quarts of the best sack [white fortified wine]: and when the ale hath done working, put these in, and stop it close six or seven days, and then bottle it, and a month after you may drink it.*

The Closet of the Eminently Learned Sir Kenelme Digby Kt. Opened, 1669

杏仁素培根
Slices of bacon imitated in marzipan

取一些杏仁膏，用红檀香染红，压成一张薄饼。另外准备一份没有染色的白杏仁膏，同样压成一张薄饼。将四层红色薄面饼和三层白色薄面饼相间叠放在一起，然后由上到下切成薄片，待薄片风干后看起来就像是培根片一样。

——《女王的乐趣》，1671

♣这道食谱在十七世纪是最受欢迎的食谱之一，在很多印刷或手写的烹饪书中都有所提及。这种用一种食物仿制另一种食物的做法可以追溯到中世纪，十六世纪随着宴会菜系的发展，这种菜变得异常流行，一般宴会的最后一道菜就会是这种仿制菜。这时宾客们通常会移步到另一个房间，或者是一个设计独特的宴会厅，在这些地方，他们往往喜欢吃一些由糖、水果和香料做成的模样精美、外形诱人的食品。在上述食谱中，将红色和白色的杏仁膏卷到一起，上下切成片状，看起来就像条状的培根片。如今在很多海边的甜品店里你仍能看到这种制作食物的方法，那里有很多用颜色各异的糖果做成的培根、鸡蛋或火腿肠等。

SLICES OF BACON IMITATED IN MARZIPAN

TO MAKE COLLOPS LIKE BACON OF MARCHPANE. *Take some of your marchpane [marzipan] Paste, and work it in red saunders [sandalwood – a red dye] till it be red; then rowl a broad sheet of white paste, and a sheet of red paste; three of the white and four of the red, and so one upon the other in mingled sorts, every red between, then cut it overthwart, till it look like collops [slices] of bacon, then dry it.*

A Queen's Delight, 1671

天鹅馅饼和三角胸衣

A great swan pie and a swan-skin stomacher

取一只天鹅，去掉羽毛后洗净。剥掉天鹅皮，再剔除骨头，只留下天鹅肉。在天鹅肉上涂上猪油，并用胡椒粉、食盐、丁香和肉豆蔻干皮调味。根据天鹅肉的多少制作黑麦面饼，用天鹅肉做馅，在面饼里层涂上一层猪油并撒上一层月桂叶，用黄油将面饼的口封住。将天鹅头和天鹅腿放在馅饼上用于装饰和点缀，然后进行烘焙，冷却后再抹上一些黄油。剥下来的天鹅皮晒干后可以用来做成三角胸衣，防止胸部受寒。

——威廉·拉比沙，《剖析烹饪全过程》，1682

在英格兰，人们最早从中世纪就开始食用天鹅肉了，虽然天鹅在那个时期也往往被看作珍贵物种。天鹅一般是用来烤着吃，而且烤完后会重新给天鹅肉“穿”上天鹅皮（参照前面烤孔雀的制作方法）。上述这一特别的食谱是将去骨的天鹅肉填充到一个大的馅饼里，然后用黄油将馅饼密封住。在这道食谱中用不到天鹅皮，因此拉比沙建议可以把剩下的天鹅皮做成时尚配饰。在这之后不久，天鹅肉就开始淡出人们的视野，在特殊场合或宴会上，逐渐被易于饲养的火鸡取代。接下来很快就会给大家介绍到火鸡的制作食谱。

A GREAT SWAN PIE AND A SWAN-SKIN STOMACHER

TO BAKE A SWAN. *Pull all the gross feathers from the swan clean, and all the down; then case [skin] your swan, and bone it, leave all the flesh, lard it [with fat] extreme well, and season it very high with pepper, salt, cloves, and mace; so having your coffin [pastry case] prepared in the proportion of the swan, make of rye dough, put in your swan, and lay some sheets of lard and bay-leaves on top, so put on butter and close it: put on the head and legs on the top, garnish and indore [gild] it, and bake it; when it is cold fill it up with clarified butter. Your skin being spread forth and dried, is good to make a stomacher for them that are apt to take cold in their breast.*

William Rabisha, The Whole Body of Cookery Dissected, 1682

土豆水果馅饼
A sweet and spicy potato and fruit pie

将西班牙土豆（不要太多）煮熟后，切成大拇指一样厚度的土豆片，用肉豆蔻、桂皮、生姜和白糖调味。取事先准备好的馅饼，将土豆片放到馅饼底部，然后加入调好味的西葫芦、一把葡萄干、枣、糖制橘子皮、糖制香橼皮和切好的糖制海冬青根，最后在上面加上黄油就可以烘焙了。用一点醋、烈性白葡萄酒、糖、一个鸡蛋黄和一点黄油制成酱汁，待馅饼做熟后倒上酱汁，撒上白糖，加上配菜，然后就可以上桌了。

——威廉·拉比沙，《剖析烹饪全过程》，1682

☘在现代社会，我们经常会用土豆做一些开胃菜。上面这道早期食谱中提到的西班牙土豆属于甜土豆还是普通土豆我们不得而知，因为不论是甜土豆还是普通土豆，多数都是经由西班牙运往欧洲的。十七世纪的食谱仍旧喜欢将甜和咸搭配在一起，即使是开胃馅饼中裹的肉馅，也会在最后一道工序时撒上一层糖霜。

A SWEET AND SPICY POTATO AND FRUIT PIE

TO MAKE A POTATO PIE. *Boyle your Spanish potatoes (not overmuch) cut them forth in slices as thick as your thumb, season them with nutmeg, cinnamon, ginger, and sugar; your coffin [pastry case] being ready, put them in, over the bottom, add to them the marrow of about three marrow-bones, seasoned as aforesaid, a handful of stoned raisons of the sun, some quartered dates, orangado [candied orange peel], cittern [candied citron peel], with ringo-roots [candied root of eringo or sea holly] sliced, put butter over it, and bake them: let their lear [a thickened sauce] be a little vinegar, sack [white fortified wine] and sugar, beaten up with the yolk of an egg, and a little drawn butter; when your pie is [cooked] enough, pour it [the sauce] in, shake it together, scrape on sugar, garnish it, and serve it up.*

William Rabisha, The Whole Body of Cookery Dissected, 1682

樱草花馅饼

A tart made with a gallon of cowslip flowers

收集至少一加仑樱草的花朵，切碎后放到研钵中捣成浆。加入一把或两把蛋白杏仁饼干粉，一品脱或半品脱奶油，搅拌后一起放到平底锅中稍微加热，然后关火。再将八个鸡蛋和一点奶油搅拌在一起，如果不够浓稠，放到温火上加热至浓稠为止，但注意不要让它凝固，然后倒入平底锅中。最后，用白糖、玫瑰花水和一点食盐调味。你可以用盘子烘焙果馅饼，但是最重要的是保证放入鸡蛋中的奶油是凉的。

——威廉·拉比沙，《剖析烹饪全过程》，1682

♣制作这道赏心悦目的果馅饼需要非常多的樱草花，可见这种花在十七世纪的英格兰比现在要更为常见。樱草花一般用作配料或装饰。日记作者兼美食作家约翰·伊夫林（John Evelyn）曾提议在沙拉中加入樱草花，而休·普拉特（Hugh Plat）在食谱中也提到将那些仍旧在田野中生长茂盛的花朵做成蜜饯。弗洛伦斯·怀特（Florence White）在1934年出版的《以花为食》（*Flowers as Food*）一书中列出了樱草奶油、糖浆、茶、葡萄酒、蜂蜜酒以及樱草蛋糕、樱草泡菜、樱草醋和一道十八世纪早期用樱草做成的蜂窝蛋糕的食谱。在埃塞克斯，人们称樱草为“派格”，将花瓣与面粉混合在一起就可以做成“派格布丁”。

A TART MADE WITH A GALLON OF COWSLIP FLOWERS

TO MAKE A COWSLIP TART. *You must take the blossoms, of at least a gallon of cowslips, mince them exceeding small, and beat them in a mortar, put to them a handful or two of grated naples-bisket [similar to sponge fingers or macaroons], about a pint and half of cream, so put them into a skillet, and let them boyle a little on the fire, take them off, and beat in eight eggs with a little cream, if it doth not thicken, lay it on the fire gently until it doth, take heed it curdles not, season it with sugar, rosewater; and a little salt; you may bake it in a dish, or little open tarts, but your best way is to let your cream be cold before you stir in your eggs.*

William Rabisha, The Whole Body of Cookery Dissected, 1682

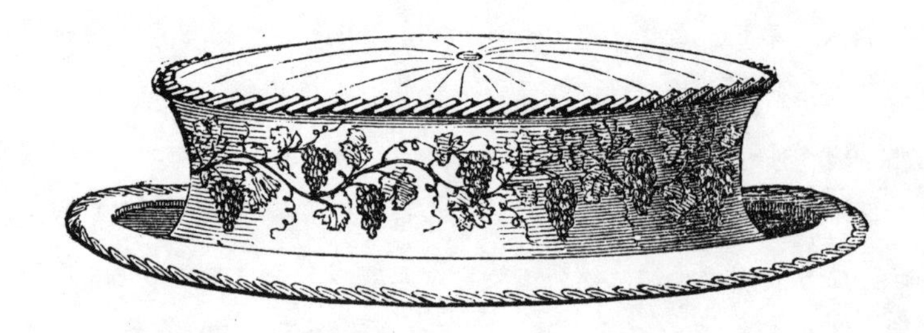

大叶白蜡子咸菜

A pickle of green ash keys

在五月收集成熟且鲜嫩的大叶白蜡子，放入一半是醋一半是水的混合液体中，再加一点食盐。液体以刚刚将白蜡子淹没为佳。放在热木炭火上加热，但不要让液体沸腾。时常搅动，白蜡子起初会变黄，继续搅动直至白蜡子变为绿色就可以停止加热了。将白蜡子铺展在木板上，完全变凉后与一块明矾一同放入装有新鲜食醋的瓦罐中，加盐，然后将瓦罐密封。

——《女王密室大公开》，1696

♣大叶白蜡子，在古典时期被医生们视为非常珍贵的药材。白蜡树的叶子可以被用来泡茶，除此之外，白蜡子是白蜡树唯一可以食用的部分。日记作者兼美食作家约翰·伊夫林在他1699年写的一本书中记录了一道与此非常类似的食谱，这本书的名字叫作《沙拉说》（*Acetaria: A Discourse of Sallets*）。还可以用马槟榔来代替白蜡子。

A PICKLE OF GREEN ASH KEYS

TO PICKLE ASHEN KEYS. *Take the youngest keys [seed bract of the ash tree] in May, when they are full-grown and tender, put them in a liquor made of half vinegar and water, and some salt; put no more upon them than what will cover them; set them upon hot embers, but let them not boyl; stir them often and they will be first yellow; keep them stirred until they be green, then take them out, and lay them abroad upon a board until they be cold, then put them up in fresh vinegar and salt, with a piece of alum. Cover the crock close, with a weight upon it.*

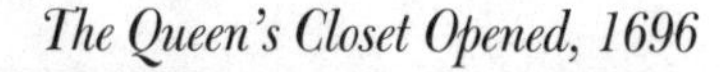

The Queen's Closet Opened, 1696

巧克力海鸭子

A sea-duck served with a sauce of chocolate, truffles and morels

将海鸭子去毛洗净后焯水。然后取出海鸭子，放到一个罐中，加入胡椒粉、食盐、月桂叶和一小把香草调味，再加入一点巧克力，炖煮。与此同时，用鸡肝、羊肚菌、小牛蘑菇、普通蘑菇、松露和一夸脱栗子做成一份菌菇汤。将海鸭子装盘，浇上做好的菌菇汤，再配上想要吃的开胃菜，就可以端上桌了。

——马西亚洛，《宫廷和乡村厨师》，1702

♣现在有很多可以被称作海鸭子的物种，包括绒鸭、海番鸭、白颊鸭和秋沙鸭。在美国南部的烹饪中经常会看到用巧克力给开胃菜调味，法国厨师马西亚洛在创造上述食谱时可能是受到了意大利烹饪的影响，因为意大利的厨师们曾在十七世纪八十年代尝试过在荤菜中加入少量的巧克力。马西亚洛的书第一次在法国出版是在十七世纪九十年代，被翻译成英文后于1702年在伦敦出版，名为《宫廷和乡村厨师》（*The Court and Country Cook*）。上述食谱好像并没有对英国的饮食造成很大的影响，直到二十世纪末墨西哥餐厅来到英国之前，巧克力在英国很少被加在开胃菜中。

A SEA-DUCK SERVED WITH A SAUCE OF CHOCOLATE, TRUFFLES AND MORELS

A SEA-DUCK WITH CHOCOLATE IN A RAGOO. *Having pick't, cleans'd and drawn your sea-duck, as before, let it be wash'd, broiled a little while upon the coals, and afterwards put in a pot; seasoning it with pepper, salt, bay-leaves and a faggot [bundle] of herbs. Then a little chocolate is to be made and added thereto; preparing at the same time a ragoo [from the French ragout, or stew] with capons-livers, morilles [morel mushrooms], mousserons [the sweetbread mushroom], common mushrooms, truffles, and a quarter of a hundred of chestnuts. When the sea-duck is ready dress'd in its proper dish, pour your ragoo upon it; garnish it with what you please, and let it be serv'd up to table.*

François Massialot, The Court and Country Cook, 1702

獾肉火腿

Badger ham anyone?

獾是世界上最干净的生物之一，没有人会说吃獾的肉会对身体不好。它的肉吃起来像上好的猪肉，但是比猪肉甘甜。将獾杀死后，切下它的后腿，去皮。然后将獾的后腿在盐水中浸泡一周或十天，盐水的浓度最好能够浮起一个鸡蛋。浸泡完后放入水中，持续加热四到五小时后取出，撒上一些面粉和面包屑，然后插上烤叉烤制。吃的时候要趁热，还可以配上炸培根和柠檬片等一同食用。

——理查德·布兰德利，《乡村主妇和妇女的导师》，1736

理查德·布兰德利（Richard Bradley）是一位牛津植物学教授，同时也是一位撰写主妇指南的作者，他曾经研究过獾肉火腿这道食谱，并且最后还将它插在制作青蛙和蜗牛的一系列法国烹饪书之后，因此，莱斯特郡[1]一位名叫R.T.的先生称理查德为“好奇宝宝”。教人制作獾肉的食谱非常少，据说獾肉吃起来像野味牛肉，最好搭配咖喱食用。然而，另一道关于獾肉的食谱也非常有趣，它在一本法国乡村食谱集中得以留存下来。食谱的开头非常巧妙：“将獾的内脏取出，皮毛剥掉，放在水流湍急的小溪中浸泡四十八小时，这可以帮助你更快地去掉獾身上的油脂。”

① 莱斯特郡（Leicestershire）是英国英格兰中部的郡。

BADGER HAM ANYONE?

A GAMMON OF A BADGER ROASTED. *The badger is one of the cleanest creatures, in its food, of any in the world, and one may suppose that the flesh of this creature is not unwholesome. It eats like the finest pork, and is much sweeter than pork. Then, just when a badger is killed, cut off the gammons, and strip them; then lay them in a brine of salt and water, that will bear an egg, for a week or ten days then boil it for four or five hours, and then roast it, strewing it with flour and rasped bread [breadcrumbs] sifted. Then put it upon a spit… Serve it hot with a garnish of bacon fry'd in cutlets, and some lemon in slices.*

Richard Bradley, The Country Housewife and Lady's Director, 1736

蝰蛇汤和蝰蛇肉汤

Viper soup, or viper and snail broth –the choice is yours

蝰蛇汤：取一条活的蝰蛇，去掉蛇头，剥皮后切成两英寸长的肉块，连着蛇心一同放到水中炖煮。如果蛇够大的话，一加仑水中放八条蝰蛇的量比较适宜。在水中加入一点胡椒粉、食盐以及一夸脱白葡萄酒，再加入一点辛香料。取山萝卜、白甜菜或菜叶、卷心菜和生菜的菜芯、葱、菠菜叶和菊苣，切碎后一同放入锅中。煮熟后，趁热卷到法式面包卷中，用面包屑和柠檬片做配菜，即可食用。

——理查德·布兰德利，《乡村主妇和妇女的导师》，1736

蝰蛇肉汤：取两条蝰蛇，去掉蛇头，剥皮后切成十六段，用肉豆蔻干皮、食盐和甜胡椒调味。依照自己的口味，将蛇肉放入两夸脱优质鱼汤或牛肉汤抑或是清水中，加热炖煮。煮到一半时可加入一把香草和一根葱。若用同样的方法做蜗牛汤，需要将蜗牛洗净、去壳，在盐水中搓洗，然后放入两夸脱水中炖煮，煮到锅中只剩一夸脱水为止，调味参照蝰蛇汤。

——查尔斯·卡特，《卓越的宫廷和乡村厨师们》，1736

不管是在过去还是现在，世界各地都有吃蛇的习惯，但是蝰蛇汤在英格兰好像只流行了一小段时间。理查德·布兰德利从一位用各种酱料和焖肉做蜗牛的法国厨师那里获得灵感创造出了蝰蛇汤这一食谱。蝰蛇肉汤与蝰蛇汤不同，蝰蛇肉汤在十六到十九世纪的很多主妇指南和药草手册中十分常见，是一种药用食谱。据说，蝰蛇肉汤不仅是一种解酸药，还是一种强效滋补的补药，通常被认为可以用来提高性欲。

VIPER SOUP, OR VIPER AND SNAIL BROTH–THE CHOICE IS YOURS

VIPER SOUP. *Take vipers, alive, and skin them, and cut off their heads; then cut them in pieces, about two inches in length, and boil them, with their hearts, in about a gallon of water to eight vipers, if they are pretty large. Put into the liquor a little pepper and salt, and a quart of white wine to a gallon of liquor; then put in some spice, to your mind, and chop the following herbs, and put into it: take some ghervill [chervil], some white beetcards [chard] or leaves, some hearts of cabbage-lettuce, a shallot, some spinach leaves, and some succory [chicory]. Boil these, and let them be tender; then serve it up hot, with a French [bread] roll in the middle, and garnish with the raspings of bread [breadcrumbs] sifted, and slices of lemon.*

Richard Bradley, The Country Housewife and Lady's Director, 1736

VIPER BROTH. *Get two vipers and cut them in sixteen pieces, but use not the heads, skin them and season with mace salt and Jamaica pepper [allspice], put to two vipers, two quarts of good fish-broth, or clear veal-broth, or water, according as you will have it strong, stew this half away, and strain it, put in a faggot [bundle] of herbs and one shallot. A way to make snail broth is with snails washed and shelled and rubbed with water and salt, then stew them from two quarts of water to one, and then strain them off and season the same as the viper broth, and it is good for consumption.*

Charles Carter, The Compleat City and Country Cook, 1736

海鲜冻

A jellied rock pool complete with fish, shells and seaweed

将鲂鱼、鳗鱼、比目鱼、牙鳕、白杨鱼和牡蛎完整地放到清水中煮制，熬出一大锅鱼汤。在深盆子的底部倒入一些鱼汤，当鱼汤完全冷却凝胶后，放上两到三个小的牡蛎壳，两到三只小龙虾，一些海藻。然后再倒入一些鱼汤，放入一排白杨鱼、鲈鱼，等其再次冷却凝胶。如此反复，直到盆子装满为止。当盆中的东西冷却变硬时，将其完整地从盆中倒出，配上柠檬、生欧芹和茴香即可食用。

——查尔斯·卡特，《卓越的宫廷和乡村厨师们》，1736

♣如果近期有人在赫斯顿·布鲁门塔尔的布雷肥鸭饭店（Heston Blumenthal's Fat Duck at Bray）用过餐的话，就会发现这道海鲜冻和饭店里有名的“海的声音”（Sound of the Sea）这道菜之间的紧密联系。布鲁门塔尔给每位用餐者提供了一个iPod用于感受大海的声音，虽然这在查尔斯·卡特时期根本无法做到，但是我们仍旧可以看出，这两道菜都是想营造出一种海边水塘的既视感。两道菜之间会存在一种紧密的关联，这并没有什么可大惊小怪的，因为布鲁门塔尔本身就是一名食物历史学家，他很早之前就开始从以前的烹饪书中寻找灵感了。他最近开的一家名为“晚餐”（Dinner）的饭店就是受到英国烹饪的启发，这家饭店的菜单中所列出的每一道菜都标明了创作灵感来源的历史时期。

扫码看饭饭之交
如何演绎海鲜冻。

A JELLIED ROCK POOL COMPLETE WITH FISH, SHELLS AND SEAWEED

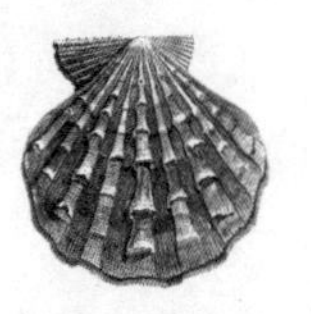

GURNETS, GUDGEONS AND OYSTERS IN ROCK JELLY. *Boil them in a good corbullion [poaching liquid], but not to pieces; let them be all whole, and make a good jelly of gurnets, eels, flounders, scate, and whiting; then put a little at the bottom of a deep bason, and when cold lay two or three small oyster-shells, and some of the sea-weed, with two or three crayfish; then some jelly, then a row of gudgeons, then perch, then jelly, till your bason is full: let it stand till cold and stiff, and turn it all out whole; garnish with lemons, raw parsly and fennel.*

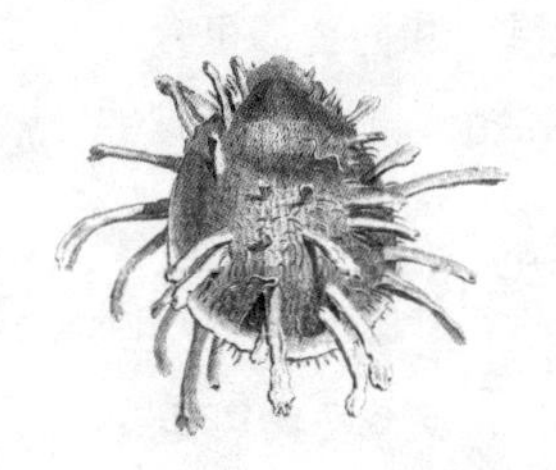

Charles Carter, The Compleat City and Country Cook, 1736

蛋壳云雀和梨形云雀

Larks, served in a nest of eggs or shaped like a pear with their feet as a stalk

蛋壳云雀：温火煮熟十二个鸡蛋或鸭蛋，对半剥开，取出里面的蛋黄和蛋白。将一半蛋壳在盘子里摆成一圈，每个蛋壳里装进一半面包屑。将云雀烤熟后，在每个蛋壳中放入一只，然后盖上另一半蛋壳，撒上金黄的面包屑。这样蛋壳云雀就做好了，它看起来和鸡蛋没什么两样。

梨形云雀：将云雀捆绑起来，去掉两只腿，用胡椒粉、食盐、丁香和肉豆蔻调味。然后用小牛胰脏、羊肚菌、蘑菇、面包碎、鸡蛋、欧芹、百里香、胡椒粉、食盐混合做成馅料，加入一点牛脂以增加馅料的黏性。参照梨的形状，用馅料包裹每一只云雀，并将云雀的腿插在上方，最后裹上一层蛋黄和面包屑，再用温火烘焙。烤好后，加上配菜即可食用。

——查尔斯·卡特，《卓越的宫廷和乡村厨师们》，1736

如今云雀是一种受保护的物种，但在过去人们非常喜欢它被烤制时特有的鲜味。现在的人都有点洁癖，甚至有点敏感，可能一想到吃云雀就会感到毛骨悚然，更何况是要吃一只被改做成梨形、上面插着细腿的云雀。十九世纪时，一个作家曾形容一只云雀去掉翅膀、腿和内脏后，剩下的残骸就像是一把牙签。

LARKS, SERVED IN A NEST OF EGGS OR SHAPED LIKE A PEAR WITH THEIR FEET AS A STALK

LARKS IN SHELLS. *Boil twelve hen or duck eggs soft; take out all the inside, making a handsome round at the top [of the shell]; then fill half the shells with passed [sieved] crumbs, and roast your larks; put one in every shell, and fill your plate with passed crumbs brown[ed]; so serve as eggs in shells.*

LARKS PEAR FASHION. *Truss your larks close, and cut off the legs and season them with pepper, salt, cloves and mace; then make a good force-meat [stuffing] with sweetbread, morelles [morel mushrooms], mushrooms, crums, egg, parsly, thyme, pepper and salt after which put in some suet and make it up stiff; then wrap up every lark in force-meat, and make it pointed like a pear and stick the leg a top; they must be washed with the yolk of an egg and crums of bread bake them gently, and serve them without sauce or they will serve for a garnish.*

Charles Carter, The Compleat City and Country Cook, 1736

黄油烤麦鹟

Baked wheatears potted in butter

麦鹟是一种来自坦布里奇的鸟。洗净后，用胡椒粉和食盐调味，放入罐中，倒上黄油，烤一个小时。一个小时后，沥干汤汁，即可食用。

——查尔斯·卡特，《卓越的宫廷和乡村厨师们》，1736

☙过去，各种各样的小鸟最终都会被乡村贫苦的老百姓们做成菜吃掉，用来补充蛋白质。如今即便在欧盟的监管下，一些欧洲南部的乡村仍旧会这么做。麦鹟是一种小型候鸟，现在被视作鹟科食虫鸟家族中的一员（这类鸟在肯特郡的坦布里奇[①]周围十分常见），它在十九世纪是一种备受人们喜爱的“小吃”。现在，人们会觉得食用这种鸣禽让人毛骨悚然，虽然鹌鹑的体形与麦鹟差不多，人们在配鹅肝酱和炸面包吃鹌鹑的时候却不会感到害怕。

奶油小牛腿

Calves feet served with a sweet currant and cream sauce

将小牛腿用温火煮至软烂后去掉牛骨头，撒上一些葡萄干，裹上半品脱奶油、两个鸡蛋黄、一点融化的黄油和白糖，趁热食用。

——查尔斯·卡特，《卓越的宫廷和乡村厨师们》，1736

① 肯特郡（Kent）是英国英格兰东南部的一个郡，坦布里奇（Tunbridge）是肯特郡下辖的一个市。

BAKED WHEATEARS POTTED IN BUTTER

POTTED WHEAT-EARS. *They are a Tunbridge bird; pick them very clean, season them with pepper and salt, put them in a pot, cover them with butter and bake them one hour; take them and put them in a cullender to drain the liquor away, then cover.*

Charles Carter, The Compleat City and Country Cook, 1736

CALVES FEET SERVED WITH A SWEET CURRANT AND CREAM SAUCE

CALF'S FEET SWEET. *You must boil them tender and take out the bones; then plump some currants, and put in half a pint of cream and the yolks of two eggs, a little melted butter and sugar, so serve away hot.*

Charles Carter, The Compleat City and Country Cook, 1736

肉馅鸡冠蘑菇汤

Little dishes of mushroom-stuffed cockscombs in gravy

将鸡冠煮至半熟，用刀尖切开，注意不要切断。准备一些家禽的肥肉、培根和牛骨髓，切碎后在研钵中碾成末，再用食盐、胡椒粉、肉豆蔻粉调味，然后加入一个鸡蛋搅拌均匀。将调好的肉馅填充到鸡冠里，放到浓肉汤中温火炖半个小时，再加入一些鲜蘑菇片和鲜蘑菇丁，然后打入一个鸡蛋黄，搅拌一下，加入些许食盐。煮熟后装入小盘中。

——汉娜·格拉斯，《简易烹饪艺术》，1747

♣鸡冠不管是作为配料还是主要食材，在很多早期食谱中都出现过。在上述食谱中，塞满了肉馅的鸡冠是作为配菜食用。查尔斯·卡特的小牛腿食谱同样也是在简单和便宜的食物外裹上甜奶油，将本身比较简单的菜品变成一道精美的菜肴。过去，大部分人对吃动物身体的任何部位都不会介意，现在我们却倾向于只吃动物的纯肉部分，而把动物的内脏和身体的其他部位统统扔掉。一个社会的经济水平越低下，人们就会越重视动物的肝脏、血液和脂肪的营养价值。在现代社会中，猎人越来越少，因此我们很难再见到屠宰的场面。最有营养价值的动物内脏，如肝脏，可能会被送到身份尊贵者的手中生吃。在食物匮乏的时代，动物脂肪会被腌制起来，内脏也会被完全利用，包括头部、蹄子等部位全部都会被做进菜里。近来，人们又重视起动物的内脏，推荐食用动物内脏的人是著有《鼻子到尾巴的猪料理》（*Nose to Tail Eating*）这本书的伦敦圣约翰饭店（London Restaurant St John）的老板费格斯·亨德森（Fergus Henderson）。

LITTLE DISHES OF MUSHROOM-STUFFED COCKSCOMBS IN GRAVY

TO FORCE COCKS-COMBS. *Parboil your cocks-combs, then open them with a point of a knife at the great end: take the white of a fowl, as much bacon and beef marrow, cut these small, and beat them fine in a marble mortar; season them with salt, pepper, and grated nutmeg, and mix it with an egg; fill the cocks-combs, and stew them in a little strong gravy softly for half an hour, then slice in some fresh mushrooms and a few pickled ones; then beat up the yolk of an egg in a little gravy, stirring it. Season with salt. When they are enough, dish them up in little dishes or plates.*

Hannah Glasse, The Art of Cookery Made Plain and Easy, 1747

海陆双拼烤串

Kebabs of ox palate, pigeon, chicken-peepers, cockscombs and oysters

将牛颚煮熟，切成两英尺长的薄片，在其中的一面点缀上一片培根。准备两三只鸽子和雏鸡，取出内脏，填充上肉馅，均匀地涂上一层猪油，再把它们按照这样的顺序穿到烤叉上：一只禽类、一片牛颚、一片鼠尾草叶、一片培根。取鸡冠、鸡睾丸和牡蛎，煮至半熟，涂上猪油，各点缀上一片培根，然后用培根和鼠尾草叶做间隔，一层层穿到烤叉上，进行烤制。准备三个鸡蛋黄、一些肉豆蔻、食盐、面包屑，在烤制的过程中，不断将它们涂到烤串上。准备两个对半切的小牛胰脏、切成四块并炸好的洋蓟，然后用葱将盘子擦一遍，将鸡和鸽子一个叠一个摆到盘子中间，其他食材摆在旁边。将一品脱优质肉汁、四分之一品脱红葡萄酒、一条鳀鱼、牡蛎汤和一块裹上面粉的黄油一同放入锅中蒸煮，再挤上一点柠檬汁，然后一同浇入盘中，用柠檬做配菜，一切妥当即可食用。

——汉娜·格拉斯，《简易烹饪艺术》，1747

♣格拉斯写的海陆双拼烤串食谱让我们想到在欧洲和美洲非常靠北的地方，像这种用烤叉在火上烤制食物的技能几近消失，这些地方在过去都会用烤叉烤制肉类、鱼类和其他食物。烤叉要靠人工翻转，有时候也会将狗拴在烤叉轮子上帮忙翻转烤叉。一名经验丰富的厨师可以把火候控制得刚刚好，烤叉离火源多远、烤的时候刷什么样的油都需要厨师掌控。烤好后还会撒上面包屑、香料，再涂上一层鸡蛋液，这样烤肉的外面就会形成一层脆壳，很多人都认为这是烤肉串最好吃的部分。烤制时让肉烧焦或让木炭灰掉到肉上的厨师都是不适合做烧烤的。值得一提的是，赫斯顿·布鲁门塔尔让人们重新认识了烤肉的这一制作方法，因为他在伦敦开的饭店“晚餐”（Dinner）里有很多富有特色的烤肉。

KEBABS OF OX PALATE, PIGEON, CHICKEN-PEEPERS, COCKSCOMBS AND OYSTERS

TO ROAST OX PALATES. *Having boiled your palates tender, blanch them, cut them into slices about two inches long, lard [insert strips of fat] half with bacon, then have ready two or three pigeons and two or three chicken-peepers [very young chickens], draw [eviscerate] them, truss them, and fill them with force-meat [stuffing]; let half of them be nicely larded, spit them on a bird-spit: spit them thus: a bird, a palate, a sage-leaf, and a piece of bacon; and so on, a bird, a palate, a sage-leaf, and a piece of bacon. Take cocks-combs and lambs-stones [testicles], parboiled and blanched, lard them with little bits of bacon, large oysters parboiled, and each one larded with a piece of bacon, put these on a skewer with a little piece of bacon and a sage-leaf between them, tie them on to a spit and roast them, then beat up the yolks of three eggs, some nutmeg, a little salt and crumbs of bread; baste them with these all the time they are a-roasting, and have ready two sweetbreads each cut in two, some artichoke-bottoms cut into four and fried, and then rub the dish with shalots: lay the birds in the middle, piled upon one another, and lay the other things all separate by themselves round about in the dish. Have ready for sauce a pint of good gravy, a quarter of a pint of red wine, an anchovy, the oyster liquor, a piece of butter rolled in flour; boil all these together and pour into the dish, with a little juice of lemon. Garnish your dish with lemon.*

Hannah Glasse, The Art of Cookery Made Plain and Easy, 1747

随身携带的即食鲜肉冻
Instant meat-stock 'glue' to carry in your pocket

取一头小牛的牛腿，剥皮并去掉肥肉脂肪，再将肉与骨头分离。将取下的瘦肉放入三或四加仑水中炖煮，要炖到锅中的肉烂了，汤浓到可以结成肉冻。注意炖煮时要保证锅盖盖牢，不要着急掀开。每隔一段时间可以用勺子舀点出来。当你觉得汤浓到足够结成冻的时候，即可将汤过滤到一个干净的砂锅中。当汤冷却后，将浮在汤上的脂肪撇掉。取一个又大又深的炖锅，锅中加水煮沸，然后取一些高瓷杯或者釉质良好的陶器，将汤冻倒入杯中，然后将杯子置于加水的炖锅中，但要确保锅中的水不会进入杯子。一直加热直至杯中的汤冻完全变成胶状物，就可以取出冷却，放在新的法兰绒布上晾六到八小时直至完全风干，然后放到干燥且温度适宜的地方。过不了多久，汤冻就会像胶块一样又干又硬，这样就可以把它们装到口袋里随身携带但又不会造成任何损坏了，不过最好是把它们放到一个小锡盒子里。想吃的时候就取出一块胡桃大小的汤冻，用一品脱水加热炖煮，不断搅拌直至完全融化后，加入食盐。如果你想加香草或香料，那要先在水中加入香草或香料煮一会儿，然后把煮过的水倒在汤冻上。

——汉娜·格拉斯，《简易烹饪艺术》，1747

♣汉娜·格拉斯的即食鲜肉冻食谱只是十八世纪和十九世纪上半叶家庭烹饪指南中众多自制即食食谱中的一道。十九世纪六七十年代，很多公司都开始在工厂生产提纯肉类，如1865年建立的李比希（Liebig）肉类提取公司。李比希在乌拉圭的弗赖本托斯设港，用欧洲肉价的三分之一购买当地养殖的牛肉。十九世纪七十年代，约翰·劳森·约翰斯顿（John Lowson Johnston）发明了“约翰斯顿液体牛肉”（Johnston’s Fluid Beef），后更名为保卫尔牛肉汁（Bovril）。与李比希肉类提纯不同的是，保卫尔牛肉汁中包含香料并且是在能供应更低价牛肉的阿根廷设厂制造。

INSTANT MEAT-STOCK 'GLUE' TO CARRY IN YOUR POCKET

TO MAKE POCKET-SOUP. *Take a leg of veal, strip off all the skin and fat, then take all the muscular or fleshy parts clean from the bones. Boil this flesh in three or four gallons of water till it comes to a strong jelly, and that the meat is good for nothing. Be sure to keep the pot close covered, and not to do too fast; take a little out in a spoon now and then, and when you find it is a good rich jelly, strain it through a sieve into a clean earthen pan. When it is cold, take off all the skin and fat from the top, then provide a large deep stew-pan with water boiling over a stove, then take some deep china-cups, or well-glazed earthen-ware, and fill these cups with the jelly, which you must take clear from the settling at the bottom, and set them in the stew-pan of water. Take great care that none of the water gets into the cups; if it does it will spoil it. Keep the water boiling gently all the time till the jelly becomes as thick as glue, take them out, and let them stand to cool, and then turn the glue out into some new coarse flannel, which draws out all the moisture, turn them in six or eight hours on fresh flannel, and so do till they are quite dry. Keep it in a dry warm place, and in a little time it will be like a dry hard piece of glue, which you may carry in your pocket without getting any harm. The best way is to put it into little tin-boxes. When you use it, boil about a pint of water, and pour it on a piece of glue about as big as a small walnut, stirring it all the time till it is melted. Season with salt to your palate; and if you chuse any herbs or spice, boil them in the water first, and then pour the water over the glue.*

Hannah Glasse, The Art of Cookery Made Plain and Easy, 1747

紫色毒梨

Hannah Glasse's poisonous purple pears

取四个梨，去皮，切块，去核，放入炖锅中，加入适量水和四分之一磅白糖，盖上锡板后再盖上锅盖，温火加热。注意观察，防止锡板融化。当汤煮成紫色后便可将梨和汤汁一同倒入盘中，冷却后可作为第二道配菜上桌，或者依你的心情上菜。

——汉娜·格拉斯，《简易烹饪艺术》，1747

有个有趣的说法，意思大概是不管怎样请不要自己在家中尝试将梨放到红葡萄酒中制作，除非你想中毒。格拉斯的食谱让酸性水果和锡板中的铅产生化学反应，制造出一种紫色的颜色变化，这种化学反应会使梨带有毒性。

扫码看视频。

HANNAH GLASSE'S POISONOUS PURPLE PEARS

TO STEW PEARS PURPLE. *Pare four pears, cut them into quarters, core them, put them into a stew-pan, with a quarter of a pint of water, a quarter of a pound of sugar, cover them with a pewter-plate, then cover the pan with the lid, and do them over a slow fire. Look at them often, for fear of melting the plate; when they are enough, and the liquor looks of a fine purple, take them off, and lay them in your dish with the liquor; when cold, serve them up for a side-dish at a second course, or just as you please.*

Hannah Glasse, The Art of Cookery Made Plain and Easy, 1747

奇异鸡蛋和仙女黄油

To cook eggs in a butter whirlpool, make one giant egg using bladders and golden fairy butter to look like vermicelli

漩涡鸡蛋：准备一个深煎锅，倒入三品脱液体黄油，加热至可以油炸食物的温度后不断搅拌，直至油锅出现一个漩涡。将鸡蛋打入锅中，黄油形成的漩涡会使鸡蛋变成圆球状。当荷包蛋炸到一般硬度的时候，取出装盘。炸鸡蛋会在半小时之后冷却、变软，这时就可以食用了。当然，你也可以将黄油换成清水进行同样的操作。

超大鸡蛋：将蛋白和蛋黄分离，分别用滤网过滤。将蛋黄倒入一个球状皮囊中，扎紧后煮至凝固，然后将煮好的蛋黄放入另一个更大的椭圆皮囊中，在两个皮囊中间灌入蛋白，扎紧后再次蒸煮。做大份沙拉的时候会用到这种做法。做炖菜的时候，也会这样一起煮五六个蛋黄，然后放到鸡蛋炖汤中。通过这种做法，你可以将鸡蛋做成任意大小。

仙女黄油：将两个水煮蛋的蛋黄放入大理石研钵中，加入一大勺橙花水，两小勺白糖粉，一同碾成糊状。然后用搅拌器将它与等量黄油一同搅拌均匀，过滤装盘。这道菜特别适合在晚餐时食用。

——汉娜·格拉斯，《简易烹饪艺术》，1747

以上三道食谱都用到了鸡蛋，这说明中世纪、都铎王朝和斯图亚特王朝时期的烹饪特色一直延续到了乔治王朝时期。这种极具创造性的烹饪不仅对烹饪技术提出了较高的要求，还需要保证烹饪时间足够充裕。同时，这几道菜也能体现汉娜·格拉斯所著烹饪书的潜在读者的特征。早期出版的那些烹饪书语言晦涩难懂，通常只适合专业厨师阅读。汉娜·格拉斯通过改编，形成了自己独特的写作风格，语言简单易懂，使其同样适合在家庭中使用。但是仍有很多菜肴制作起来十分复杂，需要准备很多配菜，只有家庭富足而且时间充裕的人才有条件制作。

TO COOK EGGS IN A BUTTER WHIRLPOOL, MAKE ONE GIANT EGG USING BLADDERS AND GOLDEN FAIRY BUTTER TO LOOK LIKE VERMICELLI

TO FRY EGGS AS ROUND AS BALLS. *Having a deep frying-pan, and three pints of clarified butter, heat it as hot as for fritters, and stir it with a stick, till it runs round like a whirlpool; then break an egg into the middle, and turn it round with your stick, till it be as hard as a poached egg; the whirling round of the butter will make it as round as a ball; then take it up with a slice, and put it into a dish before the fire; they will keep hot half an hour, and yet be soft; so you may do as many as you please. You may poach them in boiling water in the same manner.*

TO MAKE AN EGG AS BIG AS TWENTY. *Part the yolks from the whites, strain them both separate through a sieve, tie the yolks up in a bladder in the form of a ball. Boil them hard, then put this ball into another bladder, and the whites round it; tie it up oval fashion, and boil it. These are used for grand sallads. This is very pretty for a ragoo [from the French ragout, or stew], boil five or six yolks together, and lay in the middle of the ragoo of eggs; and so you may make them of any size you please.*

TO MAKE FAIRY BUTTER. *Take the yolks of two hard eggs, and beat them in a marble mortar, with a large spoonful of orange-flower water, and two tea spoonfuls of fine sugar beat to powder; beat this all together till it is a fine paste, then mix it up with about as much fresh butter out of the churn, and force it through a fine strainer full of little holes into a plate. This is a pretty thing to set off a table at supper.*

Hannah Glasse, The Art of Cookery Made Plain and Easy, 1747

用蜗牛做人造驴奶
Artificial asses' milk made with bruised snails

取两盎司珍珠大麦、两大勺苏打粉、一盎司海冬青根、一盎司土茯苓、一盎司生姜和十八只带壳蜗牛，一同放入三夸脱清水中煮沸。等三夸脱水煮至三品脱水后，加入一品脱煮沸的鲜牛奶，搅拌均匀，再加入两盎司妥卢香脂①即可。可早晚各饮用半品脱。

——汉娜·格拉斯，《简易烹饪艺术》，1747

很多草本植物志、家庭烹饪指南和烹饪书中都有人造奶的食谱。一些动物奶因具有重要的药物特性而需求量很大，但真正的动物奶又供不应求。驴奶对那些身体虚弱以及年迈的人很有好处。有趣的是，通过现代的检测显示，驴奶在脂肪量、含糖量以及蛋白质等成分上与人的乳汁十分相近，而牛奶的脂肪和蛋白质含量偏高，含糖量偏低。上述食谱中加入了带壳蜗牛，说明人们当时已经发现动物奶中含钙这一重要特点。

扫码看视频。

① 妥卢香脂（Tolu）是一种原产于南美的胶状植物香料，可用来治疗咳嗽。

ARTIFICIAL ASSES' MILK MADE WITH BRUISED SNAILS

TO MAKE ARTIFICIAL ASSES-MILK. *Take two ounces of pearl-barley, two large spoonfuls of hartshorn [ammonium carbonate, a forerunner of baking soda, made from deer antlers] shavings, one ounce of eringo [sea holly] root, one ounce of China root [edible root of the herb Smilax china], one ounce of preserved ginger, eighteen snails bruised with the shells, to be boiled in three quarts of water, till it comes to three pints, then boil a pint of new milk, mix it with the rest, and put in two ounces of balsam of Tolu [a tree resin sometimes used to treat coughs]. Take half a pint in the morning, and half a pint at night.*

Hannah Glasse, The Art of Cookery Made Plain and Easy, 1747

将一桶毛肚带到东印度群岛
A barrel of tripe to take to the East Indies

将一定量的新鲜毛肚放到容积为四加仑的木桶中，然后就可以开始腌制了：首先，倒入七夸脱加入足量食盐的泉水，盐水的浓度要能够浮起一个鸡蛋，并使鸡蛋浮出水面的部分达到一英寸（必须使用优质食盐，普通的食盐不会使食物保存太久）；然后加入一夸脱上好的白醋、两枝迷迭香和一盎司甜胡椒香料；最后，让修桶匠将木桶锁紧。在去东印度群岛的途中一定不能打开木桶，否则毛肚很容易变质。食用的时候要先把毛肚在清水中浸泡一个半小时，然后是煎是煮就由你来定了。

——汉娜·格拉斯，《简易烹饪艺术》，1747

在十八世纪的家庭烹饪指南及烹饪书中，指导如何保存食物的食谱仍旧占有很大的比重。上述食谱也能够反映大英帝国在开拓殖民地的过程中对海外各国的影响。这种影响是双向的，英国的食物和烹饪理念传播到海外，同时海外新的食谱和食物配料被引进英国。在接下来的两百多年中，很多国家尤其是印度，开始对外传播像番茄酱、酸辣酱之类的食谱，这些食谱很快被别的国家接受，并且如今看来对整个英国的饮食造成了很大的影响（可从下一道食谱中一探究竟）。但是对大英帝国的统治者们来说，有必要将毛肚和其他东西出口到东印度群岛吗？毕竟只要是有猪、牛、羊的地方就会有毛肚，在横跨热带海洋的九个月的征途中，运送任何东西都比运送一桶毛肚容易得多。

A BARREL OF TRIPE TO TAKE TO THE EAST INDIES

TO PRESERVE TRIPE TO GO TO THE EAST-INDIES. *Get a fine belly of tripe [stomach lining], quite fresh. Take a four gallon cask well hooped, lay in your tripe, and have your pickle ready made thus; take seven quarts of spring water, and put as much salt into it as will make an egg swim, that the little end of the egg may be about an inch above the water (you must take care to have the fine clear salt, for the common salt will spoil it); add a quart of the best white vinegar, two sprigs of rosemary, an ounce of allspice, pour it on your tripe; let the cooper fasten the cask down directly; when it comes to the Indies, it must not be opened till it is just a-going to be dressed; for it wont keep after the cask is opened. The way to dress it is, lay it in water half and hour; then fry or boil it as we do here.*

Hannah Glasse, The Art of Cookery Made Plain and Easy, 1747

二十年不变质的调味酱和十八世纪的伍斯特郡酱油

A ketchup that will last you twenty years and an eighteenth-century Worcestershire sauce

二十年不变质的调味酱：准备一加仑浓烈陈啤酒、一磅腌制小银鱼、一磅剥皮大葱、半盎司肉豆蔻、半盎司丁香、四分之一盎司胡椒粉、三四个大的生姜根和两夸脱大蘑菇片。将所有材料放入锅中，用文火炖煮至半稀烂状态，然后倒入一个法兰绒布包里用力挤压，等完全冷却后再次大火炖煮，这样就完成了。你可以将做好的调味酱带到印度，一勺调味酱配一磅新鲜的液体黄油，做成一种美味鱼酱，还可以用来替代肉酱。啤酒的浓度越高、年份越久，最后做出的调味酱味道就越好。

——汉娜·格拉斯，《简易烹饪艺术》，1747

妈妈酱：在一夸脱妈妈啤酒（一种用小麦芽和香草做成的啤酒）中放入四盎司小银鱼，一些肉豆蔻干皮、肉豆蔻片，一盎司丁香，半盎司黑胡椒粉，加热煮沸至剩下三分之二夸脱的啤酒，冷却后倒入瓶子中即可使用。

——伊丽莎白·拉斐尔德，《经验丰富的英国管家》，1778

♣“调味酱”（Ketchup）一词的起源颇受争议，最早可能源于中文或者马来语。两千多年前，罗马人普遍爱吃一种与调味酱非常相似的酱料，叫作鱼醋（Liquamen），这种酱料与泰国鱼酱差不多。然而，现代的调味酱很可能起源于遥远的东方，最后于十七世纪通过贸易往来和贸易扩张传播到其他国家，其中，东印度公司发挥了重要作用。伊丽莎白·拉斐尔德（Elizabeth Raffald）也写过一道调味酱食谱，这种调味酱比普通调味酱更辛辣，而且能够保存七年之久，适合带到东印度群岛。

上述两道食谱表明英国烹饪会通过使用本地的一些配料（比如啤酒）对引进的食谱进行改良，经由改良食谱做出来的酱料又会反过来通过移居在外的人们传播到其他地方。如今的超市中还有蘑菇酱的身影，但是蘑菇酱和其他各种调味酱早就被“后来居上”的番茄酱所取代。从十九世纪末开始，番茄酱就成了最受欢迎的调味酱。

A KETCHUP THAT WILL LAST YOU TWENTY YEARS AND AN EIGHTEENTH-CENTURY WORCESTERSHIRE SAUCE

TO MAKE CATCHUP TO KEEP TWENTY YEARS. *Take a gallon of strong stale beer, one pound of anchovies washed from the pickle, a pound of shalots, peeled, half an ounce of mace, half an ounce of cloves, a quarter of an ounce of whole pepper, three or four large races [dried roots] of ginger, two quarts of the large mushroom-flaps rubbed to pieces. Cover all this close, and let it simmer till it is half wasted, then strain it through a flannel-bag; let it stand till it is quite cold, then boile it. You may carry it to the Indies. A spoonful of this to a pound of fresh butter melted, makes a fine fishsauce or in the room [in place] of gravy-sauce. The stronger and staler the beer is, the better the catchup will be.*

Hannah Glasse, The Art of Cookery Made Plain and Easy, 1747

TO MAKE MUM CATCHUP. *To a quart of old mum [beer brewed from wheat malt and flavoured with herbs] put four ounces of anchovies, of mace, and nutmegs sliced, one ounce of cloves, and black pepper half an ounce, boil it till it is reduced one third; when cold bottle it for use.*

Elizabeth Raffald, The Experienced English Housekeeper, 1778

如何辨别鸟类的肉质是否新鲜
How to tell if waders, gulls and dotterels are fresh or stale

这些鸟类，肉质新鲜的时候，爪子是柔软的；不新鲜的时候，爪子是干硬的。肉比较多的话，鸟屁股会比较丰满；肉比较少的话，鸟屁股小且硬。鸟龄小的话，鸟腿比较光滑；鸟龄偏大的话，鸟腿很粗糙。

——伊莱扎·史密斯，《做卓越的家庭主妇：贵妇好伴侣》，1758

☘很多十八和十九世纪的主妇指南都面向新兴的中产阶级，那个时候的女人都倾向于做一名家庭主妇，负责料理日常事务和饮食。那时还没有食品检查和质量检测，主妇们需要学会辨别食物是否新鲜的方法。主妇指南上有很多正确识别肉、鱼的新鲜度，蔬菜和水果的成熟度，以及识别假货等小贴士。在十九世纪末之前，食品掺假十分普遍，比如，掺水的牛奶中会加入白粉，劣质面粉做成的面包里会掺入明矾甚至碾碎的骨头。

HOW TO TELL IF WADERS, GULLS AND DOTTERELS ARE FRESH OR STALE

TO CHUSE THE SHUFFLER, GODWITZ, MARREL, KNOTS, GULLS, DOTTERS, AND WHEATEARS. *These birds, when new, are limber footed; when stale, dry footed: when fat, they have a fat rump; when lean, a close and hard one; when young their legs are smooth; when old, rough.*

Eliza Smith, The Compleat Housewife: or Accomplished Gentlewoman's Companion, 1758

猪脑馅饼

Chopped brain fritters

准备好一把面包屑、一点柠檬碎皮、胡椒粉、食盐、肉豆蔻、甜墨角兰、欧芹末和三个鸡蛋黄。将猪脑煮熟，剁成碎末，与其他食材搅拌在一起。在煎锅上涂上黄油，将搅拌好的猪脑碎末放到煎锅上煎烤。如果猪脑碎末因为太稀而在煎锅上任意滑动，就可以再加一把面包屑。

——伊丽莎白·莫克森，《英国家庭烹饪示范》，1764

从这道食谱的操作步骤中很难看出这道菜的复杂程度以及需要使用的烹饪技巧。与很多早期食谱不同的是，这道食谱的配料使用十分巧妙，烹饪的步骤简单但相对比较精确，毫不含糊。在英国，食用动物器官和内脏的人越来越少，这与英国经济发展水平的提高有直接的关系，富足的英国人经常选择一些昂贵且容易制作的肉类来烹饪。

CHOPPED BRAIN FRITTERS

TO MAKE BRAIN-CAKES. *Take a handful of breadcrumbs, a little shred lemon-peel, pepper, salt, nutmeg, sweet-marjorum, parsley shred fine, and the yolks of three eggs; take the brains and skin them, boil and chop them small, so mix them all together; take a little butter in your pan when you fry them, and drop them in as you do fritters, and if they run in your pan put in a handful more of bread-crumbs.*

Elizabeth Moxon, English Housewifery Exemplified, 1764

鸡肉冻

Hens, fish, eggs, bacon, islands and playing cards in jelly

首先，用奶油和一些甜杏仁做成牛奶布丁。用巧克力将一部分布丁染成棕色，然后放到母鸡形状的模子中。再将一个水煮蛋的蛋黄碾碎，搅拌到布丁中将布丁染黄，其余的布丁保持白色不变。将七块鸡肉分别放入母鸡形状的模子，三块加入白布丁，三块加入黄布丁，一块加入棕布丁。冷却后倒入深盘中，盘子底部和周围都放上煮软切成条的柠檬皮，就像一根根的稻草一样。再在下面垫上一块牛腿冻，四周全部摆满透明果冻。这道菜在餐桌上将十分醒目。

——伊丽莎白·拉斐尔德，《经验丰富的英国管家》，1778

拉斐尔德女士有制作动物或植物形状的肉冻的嗜好，这种肉冻适合摆在餐桌中央。在上述食谱中，她将牛奶布丁做成了母鸡和小鸡的形状，一同摆在用柠檬皮条摆成的鸡窝中，四周用透明果冻围绕，仿佛清澈的湖泊一般。做这种菜的时候通常会使用燕麦布丁，但是这道菜却用杏仁和奶油做出了复杂的牛奶布丁。拉斐尔德还用牛奶布丁和透明果冻做出了有鱼的水池、海岛、所罗门神殿，甚至纸牌。她把透明的牛奶冻做成小鱼，用金色的叶子给小鱼勾边，然后将小鱼放入装有薄果冻和葡萄酒的大酒杯中，看起来就像是鱼在酒杯中游动一般。

HENS, FISH, EGGS, BACON, ISLANDS AND PLAYING CARDS IN JELLY

HENS AND CHICKENS IN JELLY. *Make some flummery [thickened milk pudding] with a deal of sweet almonds in it, colour a little of it brown with chocolate, and put it into a mould the shape of a hen; then colour some more flummery with the yolk of a hard egg beat as fine as possible, leave part of your flummery white, then fill the mould of seven chickens, three with white flummery and three with yellow, and one the colour of the hen; when they are cold turn them into a deep dish; put under and round them lemon peel, boiled tender and cut like straw, then put a little clear calf's foot jelly under them, to keep them in their places, and let it stand till it is stiff, then fill up your dish with more jelly. They are a pretty decoration for a grand table.*

Elizabeth Raffald, The Experienced English Housekeeper, 1778

惊奇兔子

Rabbits with jaw-bone horns, a bunch of myrtle in their mouths and a frothy liver sauce

取几只幼兔，串在烤叉上，腹中塞上布丁。烤制的时候，将兔子的下巴骨取下，插到兔子眼睛中，使其看起来像一对鹿角。烤完后将兔肉与兔骨分离，然后将兔肉同欧芹、柠檬皮、一盎司牛骨髓混合在一起，剁成碎末，加入一勺奶油和一点食盐。再将两个熟蛋黄和一小块黄油一起用研钵碾碎，与兔肉搅拌在一起，放到平底锅里，加热五分钟。接着，就可以把锅中的兔肉糊到兔骨上，用手将其重新捏合成一个完整的兔子形状，并用烤箱将外层烤焦。最后，做一份稠如奶油的肉汤，倒入盘中，在兔嘴里插上一撮香桃木，将兔肝煮熟，就可以一同端上桌了。

——约翰·法利，《伦敦烹饪艺术》，1800

♣在中世纪时期，只有年龄小于一岁的兔子才会被称为幼兔。直到近代早期，吃兔子肉还是一件非常奢侈的事情。十八世纪后，兔子肉就比较常见了，出现了很多养兔场，人们开始饲养兔子。上述食谱将兔子解构，在兔肉中混入牛骨髓、奶油和柠檬皮，然后以兔骨作为支架，重构成一只新兔子。这道食谱让人惊异的地方是，它将兔子的下巴骨插到兔子眼睛中模仿鹿角。因为重构的兔子肉容易垮掉，所以需要用烤箱将兔子外层烤焦来塑形。当时的烤箱是一个带有把手的坚固铁圆盘，这种圆盘可以在火上加热，菜肴在又红又烫的圆盘上不需要与火直接接触就可以被烤焦。将兔子表面烤焦需要有娴熟的技艺，对火候把控也要十分得当。当时，这种圆盘通常被用来制作焦糖布丁上的焦糖。

RABBITS WITH JAW-BONE HORNS, A BUNCH OF MYRTLE IN THEIR MOUTHS AND A FROTHY LIVER SAUCE

RABBITS SURPRISED. *Take young rabbits, skewer them, and put the same pudding into them as directed for roasted rabbits. When they be roasted, draw out the jaw-bones, and stick them in the eyes, to appear like horns. Then take off the meat clean from the bones; but the bones must be left whole. Chop the meat very fine, with a little shred parsley, some lemon-peel, an ounce of beef marrow, a spoonful of cream, and a little salt. Beat up the yolks of two eggs boiled hard, and a small piece of butter, in a marble mortar; then mix all together, and put it into a tossing pan. Having stewed it five minutes, lay it on the rabbit where you took the meat off, and put it close down with your hand, to make it appear like a whole rabbit. Then with a salamander [see opposite] brown it all over. Pour a good brown gravy, made as thick as cream, into the dish, and stick a bunch of myrtle in their mouths. Send them up to table, with their livers boiled and frothed.*

John Farley, The London Art of Cookery, 1800

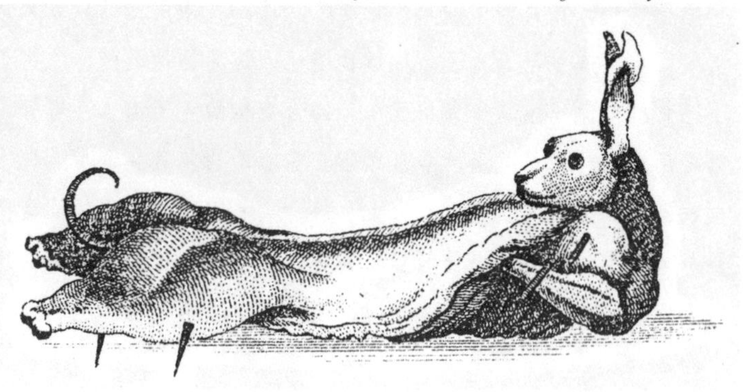

乌龟汤

How to turn a live turtle into soup and serve it in its shell

绑住乌龟的后鳍将乌龟吊起，将其头部砍下后吊一整晚。第二天早上将乌龟的前鳍切下，取出内脏，注意不要弄破乌龟胆囊。取出内脏后，切掉后鳍，并且将龟肉与龟骨分离，扯下龟壳上粘连的胶质肉，切成小片。将乌龟的前后鳍切成四英寸的长度，与胶质肉片一起洗净，放到盛有骨头和足量水的锅中，再在锅里放入一把甜香草、一个完整的洋葱，加热煮沸后，撇去表层的浮沫。当龟鳍煮熟后，取出全部龟肉，将龟汤过滤后继续加热，直至龟汤的量减少至三分之一。之后，将龟肉切成普通肉丁的四倍大小，放入另一个锅中，加入剁碎的香草混合调味料，比如墨角兰、香薄荷、百里香、欧芹、罗勒、洋葱末、香料、甜胡椒、丁香、肉豆蔻、黑胡椒、食盐和一些小牛肉汤，再将之前浓缩的龟汤倒入，继续加热。煮好后将龟肉取出，用滤布将龟汤滤出。将面粉和四分之三磅的鲜黄油加热并均匀搅拌，倒入马德拉酒（七十磅的乌龟用三品脱酒）和龟汤，沸腾后撇去浮沫，加入辣椒粉、柠檬汁和食盐调味，然后倒在切好的龟肉丁、龟壳上。如果所做的乌龟还有一些龟油，也一同将它煮熟并切成丁，加入龟肉和龟壳中，然后炖到每块肉都变软。吃的时候可以将瘦肉夹在两层肥肉中间，一同摆在盘子中，最上方再放点鸡蛋五花肉丸子和一些乌龟内脏。

——约翰·莫拉尔，《烹饪的艺术》，1808

在乔治时代后期和维多利亚时代早期，乌龟汤是最受欢迎的菜肴之一。一些大型宴会对乌龟汤的需求量非常大。在伦敦的酒馆里，乌龟汤更是一道特色菜，伦敦有很多人会专门饲养乌龟用来做乌龟汤。乌龟汤在家中不易制作，因为做菜前要保证乌龟是活着的，再加上乌龟体型较大，操作复杂。乌龟汤和用小牛肉做成的仿龟汤是最先装在罐头中出售的食物之一。最早的乌龟汤食谱出现在汉娜·格拉斯（Hannah Glasse）所著的第四版《烹饪的艺术》（*Art of Cookery*，1751）一书中，仿龟汤则在其1758年第六版的书中有所提及。

HOW TO TURN A LIVE TURTLE INTO SOUP AND SERVE IT IN ITS SHELL

HOW TO COOK REAL TURTLE. *Hang the turtle up by the hind fins, and cut off the head overnight; in the morning cut off the fore fins at the joints, and the callipee all round; then take out the entrails, and be careful not to break the gall; after which cut off the hind fins and all the meat from the bones, callipee and calipash [the gelatinous substances in the upper and lower shells of the turtle, the calipash being of a dull greenish and the calipee of a light yellow colour]; then chop the callipee and callipash into pieces; scald them together, the fins being whole, but take care not to let the scales set. When cleaned, chop the fins into pieces four inches long; wash the pieces of the callipee, callipash, and fins, and put them into a pot with the bones and a sufficient quantity of water to cover; then add a bunch of sweet herbs and whole onions, and skim it when the liquor boils. When the fins are nearly done take them out, together with the remainder of the turtle, when done, picked free from bone. Then strain the liquor and boil it down till reduced to one third part; after which cut the meat*

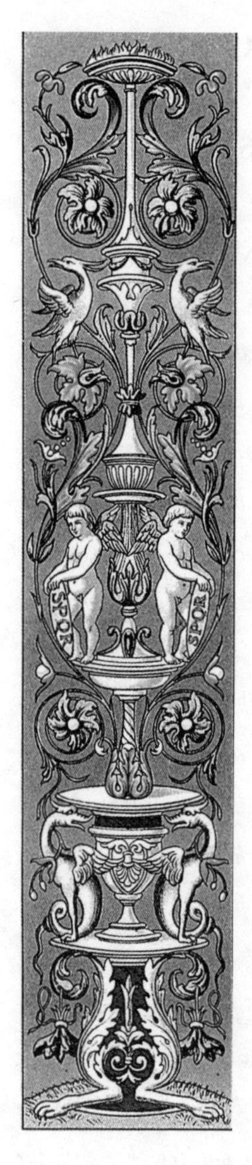

into pieces four times larger than dice; put it into a pot, add a mixture of herbs chopped fine, such as knotted marjoram, savory, thyme, parsley, a very little basil, some chopped onions, some beaten spices, as allspice, a few cloves, a little mace, black pepper, salt, some veal stock, and the liquor that was reduced. Boil the meat till three parts done, pick it free from herbs, strain the liquor through a tamis sieve, make a passing of flour and three quarters of a pound of fresh butter, mixing it well over a fire for some time, and then add to it madeira wine (if a turtle of seventy pounds weight, three pints) and the liquor of the meat. When it boils, skim it clean, season to the palate with cayenne pepper, lemon juice, and salt, and strain it to the pieces of fins and shell in one pot, and the lean meat into another; and if the turtle produce any real green fat, let it be boiled till done, then strained, cut in pieces, and added to the fins and shell, and then simmer each meat till tender. When it is to be served up, put a little fat at the bottom of the tureens, some lean in the centre, and more fat at the top, with egg and forcemeat balls, and a few entrails.

John Mollard, The Art of Cookery, 1808

情人节的毛毛心

A hairy heart for Valentine's Day

取一个小牛心脏，洗净后在心脏外面涂上一英寸厚的五香碎肉，再裹上一层米粉，然后放在盛有少量水的盘子中，放进烤箱烤制。烤熟后就着盘中的肉汤一同端上桌。这可以作为一道非常棒的配菜。

——玛丽·霍兰德，《经济的厨师和节俭的主妇》，1830

A HAIRY HEART FOR VALENTINE'S DAY

LOVE IN DISGUISE. *After well cleaning, stuff a calf's heart, cover it an inch thick with good forcemeat [stuffing], then roll it in vermicelli, put it into a dish with a little water, and send it to the oven. When done, serve it with its own gravy in the dish. This forms a pretty side dish.*

Mary Holland, The Economical Cook and Frugal Housewife, 1830

维多利亚奶酪冰激凌

A Victorian savoury cheese ice cream

将六个鸡蛋、半品脱糖水和一品脱奶油一同放入一个炖锅中，加热至黏稠状后，放入三盎司帕玛森奶酪，搅拌均匀后过滤一遍，然后冷冻。这样，冰激凌就做好了。

——伊莱扎·阿克顿，《现代家庭烹饪》，1845

♣冰激凌在十八世纪末十九世纪初颇受欢迎，那个时候伦敦的糖果贸易由很多意大利糖果商所主导，包括多梅尼科·内格里（Domenico Negri）和威廉·亚历克西斯·加林（William Alexis Jarrin）。冰激凌可做成各种各样的口味，上述冰激凌不同于其他冰激凌，它用三盎司的帕玛森奶酪做出一种甜奶油的香味，满足了大众的口味。

A VICTORIAN SAVOURY CHEESE ICE CREAM

CHEESE (PARMESAN) ICE CREAM. *Take six eggs, half a pint of syrup and a pint of cream; put them into a stewpan and boil them until it begins to thicken; then rasp three ounces of parmesan cheese; mix the whole well together and pass it through a sieve, then freeze it according to custom.*

Eliza Acton, Modern Cookery for Private Families, 1845

鸡蛋酱鳕鱼鳔

The swim bladder of a cod served with egg sauce

将鳕鱼的鱼鳔放入盐水中浸泡一晚上，第二天将褪色的皮肤组织完全去掉，洗净后放入大量的掺了水的冷牛奶中，加水温火炖煮三十到四十分钟，或者更长时间，只要不煮得太烂就好。时刻注意撇掉汤中的浮沫，否则时间长了浮沫就会沾到鱼鳔上，影响美观。最后，将煮熟的鱼鳔取出装盘，同鸡蛋酱和黄油一起食用。

——伊莱扎·阿克顿，《现代家庭烹饪》，1845

♣鳕鱼的很多部位都可以做成菜肴，包括鱼脸、鱼舌、鱼头、鱼肩和鱼卵，制作这些部位的食谱都可以找到。但是至于鳕鱼的鱼鳔，很难想象它能给人体提供什么营养，顶多就是对膀胱有一定作用吧。伊莱扎·阿克顿（Eliza Acton）十分聪敏，同时也是一个务实的烹饪作者，既然她建议配鸡蛋酱吃鳕鱼鱼鳔，那么一定是值得一试的。当然，这种做法在维多利亚时期非常适合当时朴素节俭且容易满足的家庭主妇们。

THE SWIM BLADDER OF A COD SERVED WITH EGG SAUCE

TO BOIL COD'S SOUNDS *[the floatation bladder of the fish]. Should they be highly salted, soak them for a night, and on the following day rub off entirely the discoloured skin; wash them well, lay them into plenty of cold milk and water, and boil them gently from thirty to forty minutes, or longer should they not be quite tender. Clear off the scum as it rises with great care, or it will sink and adhere to the sounds, of which the appearance will then be spoiled. Drain them well, dish them on a napkin, and send egg sauce and plain melted butter to table with them.*

Eliza Acton, Modern Cookery for Private Families, 1845

人造奶酪

The flavour and texture of mature cheese created with eggs and spices

将一些肉豆蔻干皮、肉豆蔻和桂皮碾碎，加入一加仑鲜牛奶、两夸脱奶油，加热煮沸后再加入八个鸡蛋、六或八勺葡萄酒醋，搅拌均匀后加热，直到结成凝乳。将凝乳放到干酪布中，悬挂六到八小时滤出水分。然后将其中的香料挑出，加入白糖和玫瑰水，放入滤锅中静置一小时。最后，在盘子的底部加上一层奶油，把沉淀下来的奶酪装入盘子，就可以上桌了。

——弗雷德里克·毕晓普，《伦敦插图烹饪书》，1852

♣这道食谱做出的奶酪是人造奶酪，虽然人造奶酪的香味和质地都是刻意仿造出来的，但和真正的奶酪其实相差无几。通过添加香料、鸡蛋和白糖，可以将人造奶酪做成凝乳干酪，并作为甜点供客人食用，它与生奶酪蛋糕非常相像。

THE FLAVOUR AND TEXTURE OF MATURE CHEESE CREATED WITH EGGS AND SPICES

CHEESE ARTIFICIAL. *Well pound some nutmeg, mace, and cinnamon, to which add a gallon of new milk, two quarts of cream, boil these in the milk, then put in eight eggs, six or eight spoonfuls of wine vinegar to turn the milk, let it boil till it comes to a curd, tie it up in a cheese cloth, and let it hang six or eight hours to drain, then open it, take out the spice, sweeten it with sugar and rose water, put it into a cullender, let it stand an hour more, then turn it out and serve it up in a dish with cream under it.*

Frederick Bishop, The Illustrated London Cookery Book, 1852

HENS' EGGS.

巧克力鸡尾酒

A sweet 'cocktail' of sherry, port and chocolate

取一品脱雪利酒或一个半品脱波特酒、四个半盎司巧克力、六盎司优质白糖、半盎司白淀粉或面粉，混合均匀后，加热十或十二分钟即可。如果你的巧克力里含糖，则需将用量改为九盎司巧克力和三盎司白糖。

——弗雷德里克·毕晓普，《伦敦插图烹饪书》，1852

这是被厨师赫斯顿·布卢门塔尔发现并重新改造过的又一食谱。布卢门塔尔认为至少在十七世纪末这道食谱就已经出现，因为厨师们从那个时期就开始用新兴的巧克力制作各类菜肴。布卢门塔尔的制作方法与上述食谱非常相似，不同的是他将离心机作用于巧克力和酒的混合物，使其在加入乳清蛋白粉（原食谱是加入淀粉）之后固液分离，给酒营造出一种泡沫丰富的效果。

A SWEET 'COCKTAIL' OF SHERRY, PORT AND CHOCOLATE

CHOCOLATE WINE. *Take a pint of sherry or a pint and a half of port, four ounces and a half of chocolate, six ounces of fine sugar, and half an ounce of white starch or flour; mix, dissolve, and boil these altogether for ten or twelve minutes; but if your chocolate is made with sugar, take double the quantity of chocolate and half the quantity of sugar.*

Frederick Bishop, The Illustrated London Cookery Book, 1852

北极果冻
Arctic jelly

将四盎司冰岛海苔放到温水中洗净，滤净水分，放入一夸脱清水中煮，并不断搅拌。温火煮一个小时后，加入四盎司白糖、一及耳雪利酒、两个柠檬的柠檬汁、半个柠檬皮以及用一个鸡蛋的蛋清与一及耳冷水混合成的蛋液，继续加热并不断搅拌。果冻成型后倒入法兰绒滤袋中，水分滤净后，吃起来就不会像其他果冻那么冰凉了。有必要啰唆的是，洗冰岛海苔时要注意不要将表面的奎宁洗掉，吃的时候虽然会有点苦，但是对身体是有益的。冰岛海苔在欧洲大陆经常被用来治疗肺痨，在治疗严重的咳嗽和咳痰等肺部疾病上效果也比较显著。

——查尔斯·弗兰卡泰利，《工人阶级的简易食谱》，1852

♣冰岛海苔实际上是一种青苔，过去常被用来熬汤和炖粥。一般生长在英格兰北部山岭地区、苏格兰高地，以及北欧和北美的一些更高、更寒冷的地方。冰岛海苔也会被用来做一些民间的药引，据弗兰卡泰利讲，冰岛海苔曾被用来治疗过肺痨，而且颇有成效。

ARCTIC JELLY

ICELAND MOSS JELLY. *Wash four ounces of Iceland moss in some warm water, strain off the water, and put the moss on to boil in a quart of water, stirring it on the fire until it boils; it must then be removed to the side, covered over, and allowed to simmer gently for an hour; then add four ounces of sugar, a gill of sherry, the juice of two lemons, the peel of half a lemon, and a white of egg whisked with half a gill of cold water; stir the jelly on the fire until it boils, and pour it into a flannel jellybag; when passed tolerably clear, it may be taken warm, in which state it is most beneficial, or it may be eaten cold like any other jelly. It is necessary to add, that washing the moss deprives it of its tonic powers; and it is therefore recommended to put up with the bitter taste for the sake of its benefit. Iceland moss is very generally used on the Continent in the treatment of consumption; it is most active in the cure of severe coughs, and all phlegmatic diseases of the chest.*

Charles Francatelli, A Plain Cookery Book for the Working Classes, 1852

扫码看视频。

比顿女士的慈善营养汤

Mrs Beeton's nourishing soup for the deserving poor

配料：一块牛腮肉、几块便宜的牛肉（差不多四磅）、几块牛骨头、半配克洋葱、六根韭菜、一大把香草和大约半磅芹菜（只要外面的几片和一些菜叶）、半磅胡萝卜、半磅大头菜、半磅粗红糖、半品脱啤酒、四磅普通米饭或珍珠麦、半磅食盐、一盎司黑胡椒粉、一点面包屑、十加仑水。

方法：将肉切片，牛骨敲碎，放到加有十加仑水的铜锅中炖煮半小时。将准备好的蔬菜切碎，同白糖和啤酒一起放入锅中，再炖煮四小时即可出锅。不过在炖两小时后就要加入米饭和面包屑，与汤一同搅拌均匀，小火再慢炖两小时。如果觉得锅中的汤所剩不多，可以在中途添水。

——伊莎贝拉·比顿女士，《家务手册》，1861

自从十八世纪晚期之后，专为穷人制作的慈善营养汤在很多烹饪书中都有出现。在福音派基督教会的影响下，很多女性，同简·奥斯汀书中呈现的女主人公们一样，经常会对穷人投以很多关注，从主妇指南中学做一些营养汤类或药补带给穷人，这种善举一直持续到十九世纪。1858年的冬天，伊莎贝拉·比顿女士，作为一名维多利亚时期典型的中产阶级妇女，每周都会制作八或九加仑的营养汤，分发给那些靠近她的家乡宾纳尔市（Pinner）的米德尔塞克斯（Middlesex）的村民们。她说她有理由相信“大家都非常喜欢这种营养汤，给村民分发这种可以暖身子又可以充饥的营养汤比那些冷冰冰的肉和面包要好得多。对于这些普通的村民来说，他们不需要花钱就可以每天都吃到这种做法讲究而又热乎乎的饭菜”。

MRS BEETON'S NOURISHING SOUP FOR THE DESERVING POOR

USEFUL SOUP FOR BENEVOLENT PURPOSES.

Ingredients: An ox-cheek, any pieces of trimmings of beef, which may be bought very cheaply (say 4 lbs.), a few bones, any pot-liquor the larder may furnish, ½ peck of onions, 6 leeks, a large bunch of herbs, ½ lb. of celery (the outside pieces, or green tops, do very well); ½ lb. of carrots, ½ lb. of turnips, ½ lb. of coarse brown sugar, ½ a pint of beer, 4 lbs. of common rice, or pearl barley; ½ lb. of salt, 1 oz. of black pepper, a few raspings [breadcrumbs], 10 gallons of water.

Mode: Cut up the meat in small pieces, break the bones, put them in a copper, with the 10 gallons of water, and stew for ½ an hour. Cut up the vegetables, put them in with the sugar and beer, and boil for 4 hours. Two hours before the soup is wanted, add the rice and raspings, and keep stirring till it is well mixed in the soup, which simmer gently. If the liquor reduces too much, fill up with water.

Mrs Isabella Beeton, Book of Household Management, 1861

比顿女士为孩子们做的实惠牛奶汤

Mrs Beeton's extremely economical soup for children

准备两夸脱牛奶、一勺食盐、一勺桂皮粉、三勺白糖（如果喜欢吃甜的话，可以多加）、四片薄面包和六个鸡蛋黄。在牛奶中加入食盐、桂皮粉和白糖，加热煮沸。将面包片放到一个深盘子中，在上面倒上一点牛奶，然后放到炉子上保温，注意不要烤焦。将鸡蛋黄倒入牛奶中，加热并不断搅拌，不要让其结块。直至牛奶变浓稠后，全部再倒到面包片上，即可食用。一夸脱牛奶汤花不了太多钱，也就八便士，而且这道牛奶汤全年都适合饮用，做一次可供十个孩子食用。

——伊莎贝拉·比顿女士，《家务手册》，1861

这道汤适合做给孩子们吃，一般可以在早餐或者下午茶的时候食用，但不适合用来救济贫困的穷人。维多利亚时期著名的厨师和宴会负责人亚历克西斯·索亚（Alexis Soyer）建议，照顾孩子时可以用以下食谱："在早晨八点钟的时候可以给孩子喝牛奶和吃面包，正餐可以吃这些：烤羊肉和苹果布丁、烤牛肉和葡萄干布丁、萝卜炖羊肉配米饭或粉丝布丁，偶尔还可以同羊脂水饺一起吃一点酱牛肉，什么也不加或者加一点葡萄干和豌豆布丁……下午茶的时候除了这些还可以加上涂黄油的面包、牛奶和水，晚餐可以加上面包和奶酪。"

MRS BEETON'S EXTREMELY ECONOMICAL SOUP FOR CHILDREN

MILK SOUP *(a Nice Dish for Children). 2 quarts of milk, 1 saltspoonful of salt, 1 teaspoonful of powdered cinnamon, 3 teaspoonfuls of pounded sugar, or more if liked, 4 thin slices of bread, the yolks of 6 eggs. Boil the milk with the salt, cinnamon, and sugar; lay the bread in a deep dish, pour over it a little of the milk, and keep it hot over a stove, without burning. Beat up the yolks of the eggs, add them to the milk, and stir it over the fire till it thickens. Do not let it curdle. Pour it upon the bread, and serve. Average cost, 8d. per quart. Seasonable all the year. Sufficient for 10 children.*

Mrs Isabella Beeton, Book of Household Management, 1861

比顿女士的浓缩固体汤料
Mrs Beeton's stock cubes

准备两个牛肘、三个牛胫骨、一大把香草、两片月桂叶、两根芹菜、三个洋葱、三根胡萝卜、两片肉豆蔻叶、六棵丁香、一勺食盐和足量的水。取出牛骨中的骨髓后，将所有配料放入汤锅中，加适量的水，使水刚好覆盖配料，小火慢炖十二个小时或者更久，直至肉完全炖烂。取出牛肉，滤干水分，放到一个温度低的地方，将牛肉和牛骨分离。另一边，继续加热锅中的汤，直至汤量即将耗尽，但是注意不要烧焦。不盖锅盖，大火加热八小时并不断搅拌，然后将余下的汤料倒入一个深盘子中，静置一天。隔天，准备一锅水，将装有汤料的深盘子放入其中，持续加热。不断搅拌盘子中的汤料，直到汤料彻底变黏稠。然后将其倒入一个小杯子或小盆中塑形，冷却后用法兰绒布吸去水分，最后放在一个锡桶中风干。做一份这样的汤，大概要花16先令。

——伊莎贝拉·比顿女士，《家务手册》，1861

♣这个食谱表明，家庭制作的肉类提取工艺与一百多年前汉娜·格拉斯的制作方法差别甚微（详见本书第100页）。二十世纪初期，李比希公司将他们之前研制并售卖的液体肉类提取物改为了浓缩固体汤料，并将其命名为OXO。1908年，OXO曾赞助过伦敦奥运会（当时伦敦奥运会的第一赞助商是可口可乐公司），李比希公司还为运动员供应OXO的饮料以增强体质。

MRS BEETON'S STOCK CUBES

PORTABLE SOUP. *2 knuckles of veal, 3 shins of beef, 1 large faggot [bundle] of herbs, 2 bay-leaves, 2 heads of celery, 3 onions, 3 carrots, 2 blades of mace, 6 cloves, a teaspoonful of salt, sufficient water to cover all the ingredients. Take the marrow from the bones; put all the ingredients in a stock-pot, and simmer slowly for 12 hours, or more, if the meat be not done to rags; strain it off, and put it in a very cool place; take off all the fat, reduce the liquor in a shallow pan, by setting it over a sharp fire, but be particular that it does not burn; boil it fast and uncovered for 8 hours, and keep it stirred. Put it into a deep dish, and set it by for a day. Have ready a stewpan of boiling water, place the dish in it, and keep it boiling; stir occasionally, and when the soup is thick and ropy, it is done. Form it into little cakes by pouring a small quantity on to the bottom of cups or basins; when cold, turn them out on a flannel to dry. Keep them from the air in tin canisters. Average cost of this quantity, 16s.*

Mrs Isabella Beeton, Book of Household Management, 1861

鳕鱼头和鳕鱼背

Making a meal of a cod's head and shoulders (including carving instructions)

取足量的水，淹没鳕鱼鱼身，按每加仑水加五盎司食盐的配比加入适量食盐。用盐水仔细清洗鱼身，鱼肉厚的地方和鱼肚内部可再抹点盐，清洗干净后放到盐水中腌制一到两个小时，方便入味提鲜。然后把鱼放到锅中，倒入足量的冷水。要特别注意的是，加冷水的时候不要直接将水倒在鱼身上，因为这样容易破坏鱼肉。倒入冷水后，小火慢炖。鱼汤沸腾后，如水量减少，就沿着锅边再加入一点水，同理，不要将水直接倒在鱼身上。然后加入适量的盐，小心撇去鱼汤上的浮沫，温火慢炖，直至完全炖好后，将鱼身取出，放入热盘中，用切好的柠檬片、辣根、鱼子和鱼肝装饰。这道菜可供六到八个人食用，大概花费3到6先令，而且适合在每年十一月到来年三月烹饪制作。这道菜也可以搭配牡蛎酱和黄油一起食用。

切鳕鱼头和鳕鱼背的方法：首先要在鱼身两侧的中间部分运刀，深度切至鱼骨。然后在鱼肩上划上几刀，注意不要切断。制作者需要询问宾客是否需要保留鱼子和鱼肝。注意，对鳕鱼来说，鳕鱼后背的鱼骨和鱼肩是最结实的，这两部分也是内行美食家最喜欢吃的部位。鳕鱼的鱼鳔一般在鱼脊下面，鱼鳔及连接鱼头和鱼身的胶状肉质这两部分，吃起来都非常美味。

——伊莎贝拉·比顿女士，《家务手册》，1861

♣塞缪尔·佩皮斯（Samuel Pepys）1663年1月23日于日记中提到，他和国王海军勘测员威廉·巴顿（William Batton）一起享用了鳕鱼头这道菜。十九世纪的《劳特利奇的礼仪手册》（*Routledge's*

Manual of Etiquette）中曾提到，鳕鱼头是餐桌上仅次于大菱鲆的最好吃的鱼类菜肴。自此，这道菜在后来的几百年中都广受喜爱，但是最终因为一个重要的原因，鳕鱼头和鳕鱼肩逐渐在人们的餐桌上消失了。这个重要原因就是，由于北大西洋对鳕鱼的过度捕捞以及加拿大东海岸鱼类资源的减少，入得了鱼贩们的眼并且能够端上餐桌的鳕鱼越来越少，因为鱼头的尺寸达不到要求。现如今，一个鳕鱼头和鳕鱼肩充其量只够一个人食用，两人餐的话根本不够吃，因此现代英国的菜单上出现了鳕鱼颊这道菜。在鳕鱼头和鳕鱼肩的分量恢复到足够多人一起食用之前，鳕鱼颊在很长一段时间内都会是上流社会钟情的美味佳肴。

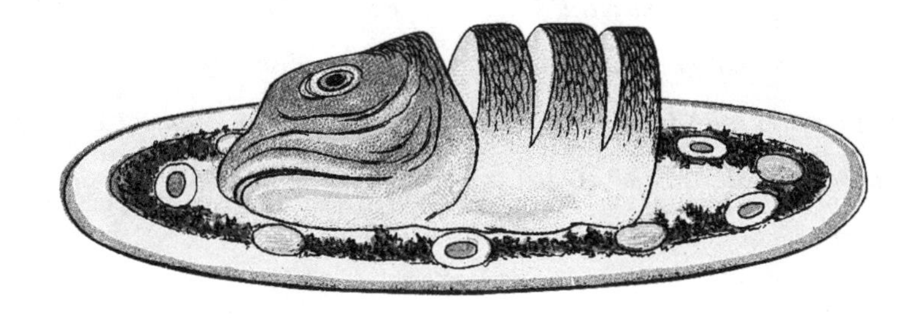

扫码看饭饭之交如何
演绎鳕鱼头和鳕鱼背。

MAKING A MEAL OF A COD'S HEAD AND SHOULDERS (INCLUDING CARVING INSTRUCTIONS)

COD'S HEAD AND SHOULDERS. *Sufficient water to cover the fish; 5 oz. of salt to each gallon of water. Cleanse the fish thoroughly, and rub a little salt over the thick part and inside of the fish, 1 or 2 hours before dressing it, as this very much improves the flavour. Lay it in the fish-kettle, with sufficient cold water to cover it. Be very particular not to pour the water on the fish, as it is liable to break it, and only keep it just simmering. If the water should boil away, add a little by pouring it in at the side of the kettle, and not on the fish. Add salt in the above proportion, and bring it gradually to a boil. Skim very carefully, draw it to the side of the fire, and let it gently simmer till done. Take it out and drain it; serve on a hot napkin, and garnish with cut lemon, horseradish, the roe and liver. Average cost, from 3s. to 6s. Sufficient for 6 or 8 persons. Seasonable from November to March. Note. Oyster sauce and plain melted butter should be served with this.*

CARVING A COD'S HEAD AND SHOULDERS. *First run the knife along the centre of the side of the fish, namely, from d to b, down to the bone; then carve it in unbroken slices downwards from d to e, or upwards from d to c, as shown in the engraving. The carver should ask the guests if they would like a portion of the roe and liver. Note. Of this fish, the parts about the backbone and shoulders are the firmest, and most esteemed by connoisseurs. The sound, which lines the fish beneath the backbone, is considered a delicacy, as are also the gelatinous parts about the head and neck.*

Mrs Isabella Beeton, Book of Household Management, 1861

军人和水手半价的便宜布丁

A cheap pudding for soldiers and sailors on half-pay

准备牛脂、红醋栗、葡萄干、面粉和面包屑各四分之一磅，两大勺糖蜜，半品脱牛奶。将牛脂剁碎，与洗净并风干的红醋栗以及无核的葡萄干、面粉、面包屑、糖蜜混合在一起，最后倒入牛奶，不断搅拌至均匀后，放到一个涂了黄油的盆中，加热蒸煮三个半小时即可。一份布丁大概花费8便士，分量可供五到六个人食用，全年可食。

——伊莎贝拉·比顿女士，《家务手册》，1861

♣一般的军官，特别是皇家海军军官，在退休或者退役之后可享受半价的优惠并领取一些津贴。简·奥斯汀（Jane Austen）在小说《劝导》（*Persuasion*）中塑造的哈维尔上校（Captain Harville）就是一个典型的例子。他与妻孩一起住在莱姆里吉斯（Lyme Regis），并没有享受什么半价优惠，只是在租来的房子中努力过活，等待着被重新召回到海上。

A CHEAP PUDDING FOR SOLDIERS AND SAILORS ON HALF-PAY

HALF-PAY PUDDING. *¼ lb. of suet, ¼ lb. of currants, ¼ lb. of raisins, ¼ lb. of flour, ¼ lb. of bread crumbs, 2 tablespoonfuls of treacle, ½ pint of milk. Chop the suet finely; mix with it the currants, which should be nicely washed and dried, the raisins, which should be stoned, the flour, bread crumbs, and treacle; moisten with the milk, beat up the ingredients until all are thoroughly mixed, put them into a buttered basin, and boil the pudding for 3-½ hours. Average cost, 8d. Sufficient for 5 or 6 persons. Seasonable at any time.*

Mrs Isabella Beeton, Book of Household Management, 1861

为病弱者做一品脱稀粥
To make a pint of gruel for invalids

准备一大勺罗宾逊牌燕麦片、两大勺冷水和一品脱开水。将去壳的燕麦片用冷水冲泡混合，然后在上面倒上开水，不断搅拌。搅拌完后倒入一个干净的炖锅中，继续搅拌并加热十分钟。最后加适量的白糖就做成了。在加热的过程中也可以用一小片柠檬皮或少量的肉豆蔻粉进行调味，但是在调味之前应该事先了解下病人的口味。最后可以把粥倒入一个平底玻璃杯中。如果病人可以喝酒的话，在粥中加入两大勺雪莉或波特酒就再好不过了。治疗体寒的话，也可以将普通的酒换成等量的烈酒。这种稀粥做一品脱的量就足够了。

——伊莎贝拉·比顿女士，《家务手册》，1861

♣简·奥斯汀在《爱玛》（*Emma*）中给大家呈现的病弱父亲伍德豪斯先生（Mr Woodhouse）大部分时间都在吃这种稀粥，幸运的话，医生也会允许他在粥中加入一勺葡萄酒。这种稀粥是燕麦粥的另外一种形式，如果将燕麦片滤出，剩下的液体变凉后就会变成果冻状。粥中除了可以加入酒或白糖外，还可以根据病人的身体状况加些猪肉丁或牛肉汁。十九世纪中后期，这种稀粥成了济贫院主要的供给食物，在狄更斯（Dickens）的小说《雾都孤儿》（*Oliver Twist*）中，奥利弗·特维斯特向济贫院乞求再给一点的就是这种稀粥，不过是不添加其他任何佐料的稀粥。

TO MAKE A PINT OF GRUEL FOR INVALIDS

TO MAKE GRUEL. *1 tablespoonful of Robinson's patent groats [oats], 2 tablespoonfuls of cold water, 1 pint of boiling water. Mix the prepared groats smoothly with the cold water in a basin; pour over them the boiling water, stirring it all the time. Put it into a very clean saucepan; boil the gruel for 10 minutes, keeping it well stirred; sweeten to taste, and serve. It may be flavoured with a small piece of lemon-peel, by boiling it in the gruel, or a little grated nutmeg may be put in; but in these matters the taste of the patient should be consulted. Pour the gruel in a tumbler and serve. When wine is allowed to the invalid, 2 tablespoonfuls of sherry or port make this preparation very nice. In cases of colds, the same quantity of spirits is sometimes added instead of wine. Sufficient to make a pint of gruel.*

Mrs Isabella Beeton, Book of Household Management, 1861

烤面包汤

How to avoid making an exceedingly disagreeable beverage

从不那么新鲜的面包上切下一片（硬的面包片做出来的效果比较好），将两面烤至金黄色，但是要注意不要烤焦。将烤好的面包片放入一个壶中，倒入开水使面包淹没在其中，然后盖紧壶盖，静置。等开水完全冷却后，就可以食用了。这种烤面包汤一定要提前做好，以备留有足够的时间冷却。如果喝的时候还是温热的，那就会特别难吃。但是凡事都有紧急的时候，如果烤面包汤要得急的话，可以在倒上开水后，不盖壶盖，直接放凉，也可以加适量的冷水加速冷却。这样做比之前的方法制作得更快。

——伊莎贝拉·比顿女士，《家务手册》，1861

烤面包汤也是适合生病体弱的人食用的一种食物，在十九世纪“饥饿的四十年代”那段时期，很多贫穷的人都是靠泡在汤中的面包来充饥，这种食物在兰开夏郡[1]被称作“博尔顿面包汤”（Bolton brewis）。十九世纪四十年代之后就不再闹饥荒了，而呈现出一种繁荣的丰收景象。制作博尔顿面包汤，首先需要将开水倒在面包屑上，随后再将水滤出，撒上食盐和胡椒粉即可。在生活水平稍好的时候，会将开水改为煮过猪肉的清汤，然后将猪肉清汤倒在燕麦饼上，等燕麦饼泡成糊状后撒上食盐和胡椒粉，单独食用或搭配黑布丁食用。

① 兰开夏郡（Lancashire）是英国英格兰西北部的郡，也是英国工业革命的发源地。

HOW TO AVOID MAKING AN EXCEEDINGLY DISAGREEABLE BEVERAGE

TO MAKE TOAST-AND-WATER. *A slice of bread, 1 quart of boiling water. Cut a slice from a stale loaf (a piece of hard crust is better than anything else for the purpose), toast it off a nice brown on every side, but do not allow it to burn or blacken. Put it into a jug, pour the boiling water over it, cover it closely, and let it remain until cold. When strained, it will be ready for use. Toast-and-water should always be made a short time before it is required, to enable it to get cold: if drunk in a tepid or lukewarm state, it is an exceedingly disagreeable beverage. If, as is sometimes the case, this drink is wanted in a hurry, put the toasted bread into a jug, and only just cover it with the boiling water; when this is cool, cold water may be added in the proportion required, the toast-and-water strained; it will then be ready for use, and is more expeditiously prepared than by the above method.*

Mrs Isabella Beeton, Book of Household Management, 1861

雪白燕窝汤

How to cook birds' nests as white as snow (pick out any small feathers first)

这种燕窝汤实际操作起来比想象中要简单得多。我在书中提到这道食谱是因为它的独创性。我想，不考虑那些稀奇古怪的构思的话，这道食谱的主要优势体现在准备过程中的清汤制作。一份汤足够十二个人食用，需要将九个燕窝在水中浸泡二十四小时，然后清洗并将粘在上面的羽毛剔除干净。随后，燕窝就会变成像面条一样的条状，此时要将仍旧粘在燕窝上的细小绒毛完全剔除。再多洗几次，直到将燕窝洗成雪一样白。将洗好的燕窝放入一个炖锅中，加入一夸脱清水，温火炖两个小时，煮好后捞出燕窝放入一个汤碗中，倒入两夸脱煮沸的清汤，即可食用。

——朱尔斯 · 顾菲，《皇家烹饪书》，1869

♣燕窝取材于东南亚地区一种常见的金丝燕的窝。这种鸟窝状如杯子，由千丝万缕交织在一起的金丝燕的唾液组成，然而也有很多人说这种燕窝的主要成分是金丝燕反刍的海藻，含有琼脂这种成分。金丝燕燕窝脱水后出口到全世界各地，也出现在了朱尔斯的日常中。燕窝是世界上最昂贵的食材之一，所以在现代社会，食谱中燕窝的用量都很小。在朱尔斯的食谱中，十二个人分食九个燕窝在如今看来是一件非常奢侈的事。

HOW TO COOK BIRDS' NESTS AS WHITE AS SNOW (PICK OUT ANY SMALL FEATHERS FIRST)

SWALLOWS' NEST SOUP. *This soup is in reality much simpler of preparation than its ambitious name would lead one to imagine. I give it here on account of its originality; and I think I may venture to say, that its principal merit, irrespective of its outlandish element, is due to the consommé employed in its preparation. For soup for 12 persons, steep, say, 9 swallows' nests in water, for twentyfour hours; wash, and pick them very carefully; the nests will then be in shreds, very similar to nouilles [noodles]; pick out any small feathers which may still adhere to the nests, and wash them again several times, until they are as white as snow; put them in a stewpan, with 1 quart of General Stock, and simmer for two hours very gently; drain, and put the nests in a soup tureen, and pour over 2 quarts of boiling consommé, prepared as directed.*

Jules Gouffé, The Royal Cookery Book, 1869

豌豆牛筋饼

Tendons with all the meat removed served in gravy with peas

取小牛胸口处的牛筋或软骨，清洗后切成两英寸长的椭圆块，放到一个炖锅中，加入足量的清汤和切碎的蔬菜，将肉块完全覆盖。加热炖煮，等到可以用针轻松地刺入其中即可。将煮好的牛筋取出，放到盘中挤压出水分，然后将肥肉去除。等完全冷却后，再去掉上面残余的牛肉，然后放到之前的肉汤中回温，最后将牛筋同绿色豌豆一起装到一个圆形大馅饼中，再在牛筋上撒上一点锅中的肉汁，即可享用。

——朱尔斯·顾菲，《皇家烹饪书》，1869

大家刚看到这道食谱的时候或许会认为，牛筋在朱尔斯所著的上流社会食谱中会是一种不上档次的菜品，这道菜本身也不常见。但是与此相类似的菜肴却在很久以前就成了意大利皮尔蒙特（Piedmont）地区的特色菜，并且因为用了整个松露做衬并配有鸡冠和舌片装饰而大受褒奖。在上述食谱中，牛筋的制作耗时长，还要挤出其中的水分，这种烹饪方法和英、法制作羊胸肉的方法十分相似。

TENDONS WITH ALL THE MEAT REMOVED SERVED IN GRAVY WITH PEAS

TENDONS OF VEAL WITH GREEN PEAS. *Cut the tendons, or gristle, from 2 breasts of veal; blanch and cut them into oval-shaped pieces 2 inches long; put them in a glazing stewpan, with sufficient consommé and mirepoix [finely chopped vegetables] to cover them entirely; simmer till the tendons are done; when tried with a trussing needle, it should enter freely; drain and press the tendons between two dishes; strain the liquor in which they have been cooked, free it of fat, and reduce it; when the tendons are cold, trim off any remaining meat which might be on them; warm them in the reduced gravy, and dish them in a circle round a croustade [pastry case]; fill the latter with green peas; pour some of the gravy over the tendons; and serve the remainder in a boat.*

Jules Gouffé, The Royal Cookery Book, 1869

袋鼠汤和咖喱，烤袋鼠和鹦鹉派

Kangaroo soup and curry, roast wallaby and parakeet pie

袋鼠尾巴汤：准备一只袋鼠尾巴、两磅牛肉、三根胡萝卜、三个洋葱、一把香草，还有一些胡椒粉、食盐、黄油和水。由关节接缝处将袋鼠尾巴切段，并放到黄油中煎至金黄。将准备好的蔬菜全部切段并煎一下。将煎好的袋鼠尾巴和蔬菜一同放到装有牛肉片和三夸脱水的炖锅中，炖煮四小时。炖熟后，取出袋鼠尾巴，滤出其他配料，然后用面粉调和汤汁使其变浓，再放入袋鼠尾巴，接着煮十分钟即可。一份可供八人食。

袋鼠尾巴咖喱：准备一只袋鼠尾巴、两盎司黄油、一大勺面粉、一大勺咖喱粉、两个洋葱的切片、一个酸苹果切丁、一小勺柠檬汁、四分之三品脱清汤以及食盐。将袋鼠尾巴洗净，擦干水分后由关节接缝处将尾巴切段，在热黄油中小火煎炸后取出。然后放入洋葱切片，油炸几分钟，不需要炸至金黄。给袋鼠尾巴裹上面粉和咖喱粉，放入炖锅中温火炖煮至少二十分钟，期间要不断搅拌。然后倒入清汤、洋葱片和酸苹果丁继续煮，同时不断搅拌。煮熟后加入一点柠檬汁或者其他调味料，盖上锅盖焖一会儿。最后装至热盘中，撒上酱料，同米饭一起食用。做这道菜从头至尾大概需要两到三小时。

烤袋鼠：准备好小袋鼠、牛肉、牛奶、黄油。由第一个关节接缝处切掉小袋鼠的后腿，给袋鼠剥皮并取出内脏，然后放在水中排出血水。将袋鼠彻底清洗干净后，把牛肉做的馅料填充到袋鼠体内并缝合起来。像烤野兔一样将袋鼠穿到烤叉上，根据袋鼠的大小，用明火烤制一个半小时到一小时四十五分钟，但是必须要和明火保持一定的距离，否则等腹中的馅料烤熟后，外面的皮肤早就烤焦了。烤制时，可以先将牛奶涂在袋鼠身上，然后再涂上黄油，再撒上面粉，涂上黄油继续烤，直至完全烤熟。

鹦鹉派：准备十二只小鹦鹉、一些煮熟的牛肉片、四片培根、三个熟鸡蛋、欧芹末、柠檬皮、胡椒粉、食盐、清汤和发酵面团。将牛肉片展开平放在一个平底锅中，上面放上六只小鹦鹉，撒上一层面粉，在空隙处填充切成片的熟鸡蛋，然后撒上调味品。将培根切成细条，和剩余的六只小鹦鹉一起码在上面，最上面码一层牛肉片，再撒一次调味品。然后在平底锅中倒满清汤或清水，最后用发酵面团封口，烘焙一个小时即可。

——伊莎贝拉·比顿女士，《家务手册》，1889

袋鼠肉是澳大利亚原住民们经常食用的肉类，在殖民主义早期，特别是十八世纪末，悉尼的市场中袋鼠肉的价格是牛肉价格的一半，这让英国殖民者很快适应了食用袋鼠肉。澳大利亚早期的一些烹饪书中有很多用袋鼠肉做食材的食谱，这些食谱后来也被比顿女士收集整理起来。比顿女士擅长制作各种形式的肉类和蔬菜以供食用，在她逝世之后这些食谱才被人们所熟知。在十九世纪末二十世纪初，沃德·洛克（Ward Locke）买下了比顿女士烹饪书的版权，她的烹饪食谱才得到广泛传播。二十世纪，袋鼠肉不再出现在澳大利亚的餐桌上了，逐渐失去了人们的喜爱，然而近几年，又重新回到了大众的视野中。

鹦鹉肉同袋鼠肉一样也是殖民主义早期澳大利亚的烹饪特色之一，多数由土著居民食用。然而，美食作家艾伦·戴维森（Alan Davidson）的报道称，有关鹦鹉和美冠鹦鹉的菜谱并不被人们看好，反而经常是人们的笑柄——“将美冠鹦鹉同一只旧鞋子一起放到足量的沸水中煮，等到鞋子煮熟后就可以将鹦鹉扔掉，食用鞋子了”。

KANGAROO SOUP AND CURRY, ROAST WALLABY AND PARAKEET PIE

SOUP FROM KANGAROO TAILS. *Ingredients: 1 tail, 2 lbs. of beef, 3 carrots, 3 onions, a bunch of herbs, pepper and salt, butter, water. Cut the tail into joints and fry brown in butter; slice the vegetables and fry them also. Put tail and vegetables into a stewpan with the meat cut in slices and boil all for four hours in 3 quarts of water. Take out the pieces of tail, strain the stock, thicken it with flour, put back the pieces of tail and boil up for another 10 minutes before serving. Sufficient for eight persons.*

CURRIED KANGAROO TAILS. *Ingredients: 1 tail, 2 oz. of butter, 1 tablespoonful of flour, 1 tablespoonful of curry-powder, 2 onions, sliced, 1 sour apple cut into dice, 1 dessert-spoonful of lemon juice, 3/4 of a pint of stock, salt. Method: Wash, blanch and dry the tail thoroughly, and divide it at the joints. Fry the tail lightly in hot butter, take it up, put in the sliced onions, and fry them for a few minutes without browning. Sprinkle in the flour and curry-powder, and cook gently for at least 20 minutes, stirring frequently. Add the stock, bring to the boil, stirring meanwhile, and replace the tail in the stewpan. Cover closely, and cook gently until tender, then add the lemon-juice and more seasoning if necessary. Arrange the pieces of tail on a hot dish, strain the sauce over, and serve with boiled rice. Time: from 2 to 3 hours.*

ROAST WALLABY. *Ingredients: wallaby, forcemeat [stuffing], milk, butter. Cut off the hind legs at the first joints and after skinning and paunching, let it lie in water for a little to draw out the blood. Make a good veal forcemeat and after well washing the inside of the wallaby, stuff it and sew it up. Truss as a hare and roast before a bright clear fire from 1 ½ to 1 ¾ hours, according to size. It must be kept some distance from the fire when first put down, or the outside will be too dry before the inside is done. Baste well, first with milk and then with butter and when nearly done dredge with flour and baste again with butter until nicely frothed.*

PARROT PIE. *Ingredients: 1 dozen paraquets (small parrots), a few slices of cooked beef, 4 rashers of bacon, 3 hard-boiled eggs, minced parsley and lemon-peel, pepper and salt, stock, puff-paste. Line a pie-dish with the beef cut into slices, over them place 6 of the paraquets, dredge with flour, fill up the spaces with egg cut in slices and scatter over the seasoning. Next put the bacon, cut in small strips, then 6 paraquets and fill up with the beef, seasoning as well. Pour in stock or water to nearly fill the dish, cover with puff-pastry and bake for 1 hour.*

Mrs Isabella Beeton, Book of Household Management, 1889

稀奇古怪的菜名

On the naming of recipes

马炮布丁：准备四分之三磅小葡萄干、半磅切好的牛脂、四分之一个切好的佛手柑、一点肉豆蔻粉、两盎司白糖、半勺食盐、一勺面粉、三个打好的鸡蛋以及一杯白兰地。将所有食材混合在一起至少煮制七个小时。注意，要在使布丁成型的模具内部涂上一层黄油。一般提前一天做好布丁，要食用的时候直接加热即可。

炮兵部队酱料：准备一杯白兰地、四分之一磅黄油、四分之一磅白糖和一个鸡蛋黄，充分混合搅拌到一起，就完成了。

兔子的拥抱：制作一些薄饼，撒上一点食盐，烘焙的时候在每个薄饼里卷上一点羊肉馅或牛肉馅，烤熟后趁热上桌。可选择搭配浓肉汁一起食用。

——《德奥利收集的罕见食谱》，1912

♣很多食谱都会起一些稀奇古怪的名字，听起来十分吸引人，光从名字也看不出它的配料，有些食谱也只是旧食谱换了新名字而已。上述三道1912年的食谱看起来与皇家炮兵、布丁或者酱料没有任何关系，而且也不知道将羊肉馅卷在薄饼里为什么要叫作“兔子的拥抱”。

ON THE NAMING OF RECIPES

HORSE ARTILLERY PUDDING. *¾ lb sultanas, ½ lb suet chopped fine, ¼ citron chopped fine, a little nutmeg grated, 2 oz sugar, ½ teaspoon salt, 1 teaspoon flour, 3 eggs well beaten, 1 glass brandy. Butter well the mould and boil for 7 hours, not less. The pudding can always be boiled the day before, then heat up next day for use.*

GUN TROOP SAUCE FOR PUDDINGS. *One glass brandy, ¼ lb butter, ¼ lb sugar, yolk of an egg. All well beaten up together.*

BUNNY HUGS. *Make some pancakes, flavoured with salt. When cooked roll into each some minced mutton or beef, and send to table very hot. Serve thick gravy either round them or separately.*

Rare Recipes, 'Collected by Deolee', 1912

牛肉杯子蛋糕
An economical savoury trifle

准备一磅切好的冻牛肉、一大勺碾碎的辣根、三盎司面包屑、一半切碎的洋葱、两盎司人造黄油、一个鸡蛋以及胡椒粉和食盐。将所有配料混合在一起搅拌，装到涂有油脂的小杯子中，温火烘焙二十分钟，烤好后放到热盘中，四周浇上肉汁即可。

——弗洛伦斯·佩蒂，《布丁夫人的烹饪书》，1917

AN ECONOMICAL SAVOURY TRIFLE

BEEF TRIFLE. *1 lb cold meat, chopped finely; 1 tablespoon horseradish, grated; 3 oz breadcrumbs; ½ onion, chopped; 2 oz margarine; 1 egg; pepper and salt to taste. Mix well together. Place in small cups, greased. Bake in moderate oven 20 minutes. Turn out on a hot dish with gravy round.*

Florence Petty, The 'Pudding Lady's' Recipe Book, 1917

扫码看饭饭之交如何
演绎牛肉杯子蛋糕。

工人阶级的茶和爵士时代的鸡尾酒

The working-class tea and jazz-age cocktails

工人阶级的茶：英国工人阶级喝的茶十分古怪，对很多美食家来说可能是极具杀伤力的一种饮品。因为工人阶级不管是吃腌肉还是鱼干，比如咸牛肉、腌鱼、腌熏鲱鱼、熏青鱼，又或者是田螺、虾、咸菜、豆瓣菜、黄瓜、莴苣、果酱、橘子酱、黄油面包、蛋糕时，都会配上这种茶。毫无疑问，这种不太搭的饮食搭配虽然对工人阶级们来说十分美味可口，但是对那些精致的享乐主义者们来说却难以接受。

胸部爱抚者：将一个新鲜鸡蛋打入一只碗中，加入四分之一品脱的鲜牛奶、两勺覆盆子糖浆、两酒杯烈性白兰地、一杯柑桂酒以及三到四个冰块，搅拌好后倒入一个小的平底玻璃杯中即可。

爱的兴奋剂：在一个雪利酒酒杯中倒入一小杯烈性白兰地以及一个新鲜鸡蛋黄、两杯樱桃白兰地和两杯黑樱桃酒，搅拌后即完成（这个酒应该一口闷）。

——J·雷伊，《用餐的全部艺术》，1921

以上这个酒席手册写于第一次世界大战结束的三年之后，在那个时期，成千上万的欧洲工人阶级都被屠杀。之后英国首相劳埃德·乔治（Lloyd George）承诺创建一个“英雄家园”，这个想法震惊了很多人。雷伊的《用餐的全部艺术》一书的写作目的是让家庭的男女主人们了解到当时底层阶级们的饮食习惯和传统。除了工人阶级之外，这本书还包括一些非常有名的爵士时代的酒类制作方法，比如上面介绍的两种名字非常露骨的鸡尾酒。

THE WORKING-CLASS TEA AND JAZZ-AGE COCKTAILS

THE TEA OF THE ENGLISH WORKING CLASS *is the most eccentric of meals, and one of the greatest injuries a gourmet could possibly conceive … for with the tea they partake of various kinds of salted meat and dried fish, such as 'corned-beef', kippers, bloaters, red herrings, winkles, shrimps, pickles, watercresses, cucumber, lettuce, jam or marmalade, bread and butter, and cake. This incongruous kind of food may, no doubt, be quite nice and tasty for this class of people, but it must shock anyone endowed with refined epicurean instinct.*

BOSOM CARESSER. *Beat up a new-laid egg into a bowl, add a quarter of a pint of fresh milk, two spoonfuls of raspberry syrup, two liqueur-glasses of liqueur brandy, a glass of Curaçao, and three or four lumps of ice. Stir well and strain into a small tumbler.*

LOVE'S REVIVER. *Put into a sherry-glass a liqueur-glass of liqueur brandy, the yolk of a new-laid egg, two dashes of Kirsch, two dashes of maraschino and serve. (This should be drunk at a gulp.)*

J. Rey, The Whole Art of Dining, 1921

叫花刺猬

Hedgehog cooked in a ball of clay

取出内脏后将刺猬清洗干净，然后将其用厚厚的一层黏土包裹起来，放到烤箱中温火烘焙，等外层的黏土变干、变硬后取出，用棍棒或锤子将黏土敲碎，刺猬的刺和外层皮肤会粘在黏土上，并且会随着黏土一起被剥下，最后再小心地剥下刺猬腿上的皮。这样，叫花刺猬就做好了。刺猬在经过上述处理后，可以使用不同方法进行烹饪，比如像做鸡肉那样烧烤或油炸，像做兔肉那样炖煮，或者像做成野兔酱那样做成刺猬酱等。

——莫杜伊子爵，《他们不会配给这些》，1940

现在大家都非常喜欢以野生动物为食，很多饭店都会从大自然中取材来为客人制作各类菜肴。第二次世界大战期间，很多人也不得不通过食用野味来应对粮食分配的不足，如此得以补充蛋白质。这种刺猬的烹饪方式有时候会被称为“吉卜赛式烹饪法”，早在二十世纪四十年代就有记载，而且也被收录在童子军们的生存手册中。

HEDGEHOG COOKED IN A BALL OF CLAY

WAYS OF COOKING A HEDGEHOG. *After killing the hedgehog, clean its inside. Then roll it thickly in some moist clay and put it in a moderate oven (Romany or kitchen oven) and bake till the clay is quite dry and hard. Break the clay by cracking it with a stick or hammer and the prickles and the top skin, which will adhere to the clay, will come off with it. Lastly, skin the legs. Having prepared the hedgehog as above, it can be cooked in different ways: either roasted or fried as a chicken, stewed as a rabbit, or made into paté as for hare paté.*

Vicomte de Mauduit, They Can't Ration These, 1940

松鼠宴

Several ways to cook squirrel, including a squirrel-tail soup

松鼠是另外一种美味，松鼠肉的味道比鸡肉更加鲜美并且更加软嫩可口。

铁板烤松鼠：将松鼠剥皮洗净，取出内脏，然后像做铁板鸡那样烤炙即可。

松鼠尾巴汤：将松鼠尾巴剥皮后，与扁豆、洋葱和香草一起煮成一份可口的尾巴汤。

烤松鼠：松鼠烤着来吃也非常美味，操作方法可参考烤鸡的食谱。

——莫杜伊子爵，《他们不会配给这些》，1940

SEVERAL WAYS TO COOK SQUIRREL, INCLUDING A SQUIRREL-TAIL SOUP

THE SQUIRREL. *This is another delicacy, the flesh of a squirrel being more tasty and tender than that of a chicken.*

GRILLED SQUIRREL. *Skin and clean the squirrel, then open it out as you would a chicken for grilling, and grill the squirrel in the same way.*

SQUIRREL-TAIL SOUP. *The tail, which is put aside after skinning, can be used with haricot beans, onions, and herbs to make a delicious soup.*

ROAST SQUIRREL. *Squirrel is also most tasty roasted, and this is done in the same way as for roast chicken.*

Vicomte de Mauduit, They Can't Ration These, 1940

炖乌鸦

How to avoid a bitter rook stew

将乌鸦剥皮并取出内脏，洗净后在其脊柱两侧分别切掉半英寸的肉，因为这部分的肉是苦的。将处理好的乌鸦放到一个砂锅中，倒入等量的清水和牛奶，使其能够覆盖住乌鸦，然后加入食盐、胡椒粉、一个切片洋葱、两棵切片大头菜、两根切片胡萝卜、一些切好的薄荷或者茴香，盖上盖子后加热炖煮即可。

——莫杜伊子爵，《他们不会配给这些》，1940

♣除了教你烹饪野生植物、水果或蔬菜外，这位美食作家在《他们不会配给这些》一书中还提到了中世纪时期很多烹饪野生鸟类的食谱，这些食谱可以让我们了解到中世纪时期英国的烹饪特点。在饥不择食的时候，只要能够获取营养和蛋白质，即便是野生鸟类，人们也会选择做来吃。在食物短缺的时代，不管是中世纪的奴隶还是二十世纪四十年代等待粮食分配的普通家庭，对野生食物胆小而又敏感的人是没有生存之地的。

HOW TO AVOID A BITTER ROOK STEW

STEWED ROOKS. *Clean, draw [eviscerate], and skin the rooks. Make an incision half an inch thick on each side of the spine and remove this piece which is the bitter part of the rook. Put the birds in a casserole with equal parts of water and milk sufficient to cover the rooks, add salt, pepper, 1 sliced onion, 2 sliced turnips, 2 sliced carrots, some chopped mint or preferably chopped fennel, and stew with the lid on the pan until tender.*

Vicomte de Mauduit, They Can't Ration These, 1940

烤麻雀和烤八哥

Vine-leaf wrapped sparrows or starlings on toast

烤麻雀：用这种方法烤麻雀可能看起来会仁慈一点。首先将麻雀的头、脖子和腿切掉，从脖子处将内脏取出，然后像烤鸽子那样给麻雀插上烤叉，在麻雀胸脯上盖上培根片，然后用葡萄叶完全包住麻雀并用线缠绕绑紧。烧烤十五分钟，并不断在上面涂油。在烧烤的过程中可以将麻雀的肝脏切碎，煎炸后浇上肉汁并调味，调好后倒在烤吐司片上。最后将烤好的麻雀放到吐司片上，再浇一次肉汁即可。

烤八哥：将八哥剥皮洗净后，像烤麻雀那样操作即可。

——莫杜伊子爵，《他们不会配给这些》，1940

VINE-LEAF WRAPPED SPARROWS OR STARLINGS ON TOAST

ROAST SPARROWS. *Sparrows when roasted in this way are far from despicable. Pluck the birds and cut off the head and neck and feet. Draw [eviscerate] the birds from the neck end, then truss them as you would pigeons, cover breasts with slices of bacon, then (if available) wrap the bird in vine leaves and tie round with string. Roast for 15 minutes, basting frequently. During this time chop finely the birds' livers, fry them in a little of the birds' gravy, season to taste, spread thickly over pieces of fried toast, place one bird on each piece of toast and pour the gravy over all.*

ROAST STARLINGS. *After skinning the birds, clean them and roast them as you would sparrows.*

Vicomte de Mauduit, They Can't Ration These, 1940

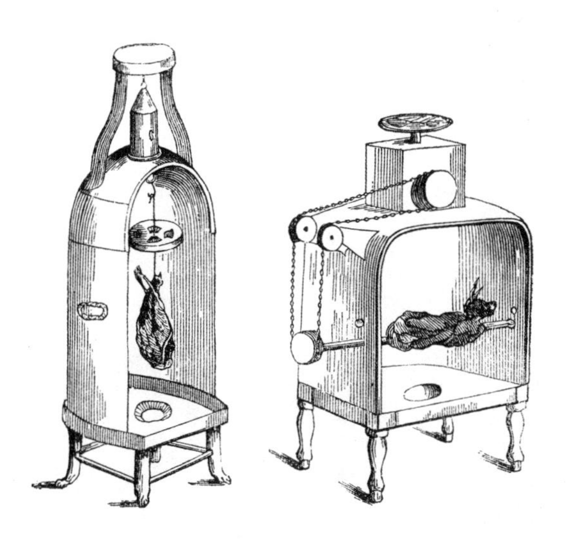

芦笋咖啡

Freshly roasted asparagus coffee

收集长熟的芦笋，洗净擦干后放在咖啡烘焙机或烤箱中烘焙，然后用咖啡研磨机或普通研钵将其碾碎，按照制作普通咖啡的步骤制作就好。这样做出来的芦苇咖啡有一种焦糖的气味，在味道上能够为大众所接受。三勺芦笋咖啡里加一勺烘焙磨碎的菊苣根粉，味道也十分不错。

——莫杜伊子爵，《他们不会配给这些》，1940

♣在拿破仑战争时期，进口咖啡豆极其短缺，拿破仑开出了很丰厚的条件以征得咖啡的替代品，上述这个芦笋咖啡就是咖啡替代品中的获胜者。这道食谱的作者从他的曾祖父那里听说了拿破仑的想法，因为他的曾祖父曾陪同拿破仑一起被流放到圣赫勒拿岛[1]。与此相类似的是，由于食物短缺，拿破仑曾经鼓励过科学家们寻求一种能够替代蔗糖的产品，这一问题最终因甜菜成功地得到了解决。

---

① 圣赫勒拿岛（Saint Helena）是南大西洋中的一个火山岛，隶属于英国。拿破仑就是被流放到这个岛上直到去世。

FRESHLY ROASTED ASPARAGUS COFFEE

COFFEE SUBSTITUTE. *Gather the berries of the grown asparagus plant, dry them, and as and when required roast them in a coffee-roaster or in a hot oven. Grind them finely in a coffee-mill or pound them in a mortar and make coffee in the usual way. This nectar has a caramel flavour and is very agreeable to the taste. Roasted and ground chicory roots can be also added in the proportion of 1 teaspoonful for every 3 tablespoonfuls.*

Vicomte de Mauduit, They Can't Ration These, 1940

厨房大作战
A kitchen goes to war

阿加莎·克里斯蒂的神奇土豆：准备六个大小适中的土豆、一点人造黄油、四勺奶油和十条小银鱼。将土豆放在烤箱中用温火烘焙，然后分别一切为二，将土豆泥挖出后与黄油、奶油一起混合捣碎，同时将小银鱼切碎混入其中，加入食盐和胡椒粉调味。最后将调好的土豆泥再装进之前的土豆壳中，在表面轻轻刷上一层人造黄油，放到热烤箱中烤至金黄即可。

奶酪辣酱饼干：杰克·沃纳说，“这是一种既美味可口又简单易做的小吃。”准备十二块薄脆饼干、两盎司人造黄油或普通黄油、两勺酸辣酱和四分之一磅奶酪。在薄脆饼干上涂一层人造黄油或普通黄油，然后将切好的奶酪片贴在上面，最后刷上酸辣酱即可。

肯尼斯·克拉克爵士的火腿卷沙拉：准备四片无骨的熟火腿片（薄切）、一份鹅肝酱、一些法式芥末、一个莴苣和一把西洋菜。将鹅肝酱涂抹到火腿片上，然后再涂一点法式芥末，像卷瑞士卷那样卷到一起，最后装到铺有莴苣片的盘中，用西洋菜的嫩叶做装饰即可。

——《厨房大作战：定额配给时期的150个名人食谱》，1940

♣在政府出版的《厨房大作战：定额配给时期的150个名人食谱》一书中，名人们被请来向大家各推荐一道经济实惠、可操作性强的食谱用以帮助度过困难的战争时期，并告诉大家“我们风雨同舟”。其中的很多食谱都是国会议员以及政府官员的妻子们推荐的，但是除此之外还有一些更加有趣的食谱，至少有些食谱的命名比内容要有趣得多。可惜的是，在“阿加莎·克里斯蒂的神奇土豆”这道食谱中，唯一神奇的地方非这道食谱的名字莫属了。除此之外，音乐厅及电台明星肯尼斯·克拉克先生（Kenneth Clark）的“火腿卷沙拉”应该是食谱书籍中最基本、最简单的食谱之一了。与前两道食谱不同的是，在当时还是伦敦国家美术馆主管的肯尼斯·克拉克先生竟然建议用鹅肝酱作为食材，这再可笑不过了，很多人对此提出了异议：“噢，朋友，难道你不知道现在在打仗吗？”

A KITCHEN GOES TO WAR

AGATHA CHRISTIE'S MYSTERY POTATOES. *6 goodsized potatoes, a little margarine, 4 tablespoonsful cream, 10 anchovies. Bake the potatoes in a moderate oven. Then cut them in half, remove the insides and mash them with the margarine and cream. Chop up the anchovies and mix them in. Add pepper and salt to taste. Return mixture to the empty skins, dap on top with margarine and brown in a hot oven.*

CHEESE AND CHUTNEY BISCUITS. *Jack Warner says: 'This may not be a "rill mill", but it is a quick, satisfying snack.' Ingredients: 1 dozen water biscuits, 2 oz. margarine or butter, 2 tablespoonsful chutney, about ¼ lb cheese. Method: Spread the biscuits with margarine or butter. Cover with cheese in slices and chutney.*

SIR KENNETH CLARK'S HAM ROLL SALAD. *4 slices of boneless cooked ham (cut thinly), 1 jar paté de foie paste, French mustard, 1 lettuce, 1 bunch watercress. Spread the paté on the cooked ham and then a little French mustard. Roll up like a Swiss roll. Dish on a bed of lettuce and garnish with sprigs of watercress.*

A Kitchen Goes to War: Famous People Contribute 150 Recipes to a Ration-time Cookery Book, 1940

土豆皮特和胡萝卜博士

Potato Pete and Doctor Carrot

杰克跳跳箱：准备六个中等大小的土豆（削好皮）、十二条鲱鱼以及煮熟的卷心菜。在每个土豆中间都用去核器纵向挖一个小洞。小洞的长度刚好可以放进两条鲱鱼，两条鲱鱼在土豆中首尾相连。用一般的烘焙方法将塞有鲱鱼的土豆烤熟，然后放进铺有卷心菜的盘中，再加上一点醋、肉豆蔻粉以及胡椒粉即可。

——粮食部，《土豆皮特的食谱》，1941

在战争时期，政府发明了土豆皮特来鼓励整个国家的人民在粮食定量发放的情况下尽可能地利用好本土种植的蔬菜。在这种高明的指示下，不管是吃面包、烤饼、调味菜还是甜品，每个人的每一餐几乎都会涉及土豆。那个时期的很多食谱都会起一个比较委婉的名字来掩盖这一事实。“杰克跳跳箱”这道菜其实就是将鲱鱼插到土豆中，然后再放到加了醋和酱料的卷心菜上。没有黄油，没有奶酪，更没有肉，有的只是可爱健康、同时能冒出鱼眼睛的干土豆，对此你又不能抱怨。除了土豆皮特之外，粮食部又想出了胡萝卜博士，宣扬胡萝卜能够提高人们在黑暗中的视力，借此可以在黑暗中躲避德军。其实，英国人在黑暗中的行动是通过雷达探测的呀。

POTATO PETE AND DOCTOR CARROT

JACK-IN-THE-BOX. *6 medium well-scrubbed potatoes, 12 sprats, cooked cabbage. Make a hole lengthwise through the centre of each potato, using an apple corer. Allowing 2 sprats for each potato, make one head emerge from one side and the tail of the second fish appear from the other end of the tunnel. Bake in the usual manner and serve on a bed of cabbage dressed with a little vinegar, a grating of nutmeg and a good shake of pepper.*

Ministry of Food, Potato Pete's Recipe Book, c. 1941

冷粥不为人知的美味
The hidden delights of cold porridge

柠檬酱三明治：用粗粮面包做成的甜三明治虽然与我们经常吃的海绵三明治不太一样，但也十分美味。用粗粮面包做三明治时，可以涂抹以下一种或几种调味品。

1. 用优质燕麦片做粥，然后与奶油蛋羹混合，增加浓稠度，最后用柠檬香精调味。

2. 制作一份浓稠的奶油蛋羹，然后在烤箱中烤制一些粗粮麦片，将烤好并冷却的麦片混入柠檬香精调味的奶油蛋羹中，混合比例由你的喜好决定。混合好后你会发现，把它涂抹在三明治上会增添一种类似果仁的美味。

——《更多厨房作战食谱》，1942

第二次世界大战期间，英国政府倡导一系列的替代食谱，用这些食谱做出的食物可以代替和平时期常吃的菜肴，而且它们还具有相同或更多的营养价值。比如，用美味的蔬菜卷、鱼、土豆薯条以及抹了人造柠檬酱的三明治做成的维多利亚蛋糕可以代替烤牛肉和约克郡布丁。这种食谱真正的可怕之处可以从所用的柠檬酱上看出来，因为柠檬酱竟然也是仿造的。

THE HIDDEN DELIGHTS OF COLD PORRIDGE

LEMON CURD SANDWICHES. *Sweet sandwiches made with bread do not seem quite the same as our old friend Sponge Sandwich, but they are very good. Make your sandwiches of national wheatmeal bread and spread them with one or other of these mixtures.*

1. Make some porridge with fine oatmeal, and use this to thicken custard flavoured with lemon essence.

2. Make a very thick custard, and meanwhile toast some coarse oatmeal in the oven. When it is done, let it get cold and then mix it with the lemon-flavoured custard in whatever proportion you like. You will find it gives the 'spread' a very pleasantly 'nutty' flavour.

More Kitchen Front Recipes, 1942

对冷粥的进一步开发利用
Further uses for cold porridge

惊奇奶油配燕麦酥：准备好两茶杯浓稠的冷粥、半罐脱水牛奶、四分之一盎司明胶、两勺细白砂糖、香草香精和两盎司无核葡萄干，喜欢的话还可以加点着色剂。再准备一盎司面粉、一盎司燕麦片、四分之一盎司猪油、一勺细白砂糖、一点塔塔粉、四分之三勺发酵粉、一撮盐以及一勺牛奶，用来制作燕麦酥。准备好配料后，就可以开始制作了。加热并不断搅拌脱水牛奶至浓稠状，然后倒入冷粥中。将明胶溶解于热水中，然后与细白砂糖、香草香精和着色剂一起倒入冷粥里，搅拌均匀后倒入一个玻璃盘中。上菜时可用葡萄干做装饰，并与燕麦酥搭配食用。燕麦酥的制作方法也十分简单，首先将所有配料混合后倒入牛奶中，调和到硬度适中后切成长条，然后放入涂过油的烘焙罐，再将烘焙罐放到热烤炉中温火烘焙，等表面呈现轻微的金黄色就表示烤好了。

——苏格兰教育部，《女王的菜肴：苏格兰重要烹饪大赛中专为女生设计的土豆燕麦饭》，1944

♣第二次世界大战期间，为了鼓励大家多食用本土种植的农作物，苏格兰教育部在食谱比赛中试图寻找一道以燕麦或土豆为主要食材的最佳食谱。一个来自基彭公立学校的女学生玛格丽特·莫尔（Margaret More）在两万名参赛者中脱颖而出，凭借以上这道食谱获得了斯特灵县（Stirling County）的冠军。让人称奇的是，在她所做的不太讲究的食物中，所用的“惊奇”的配料只是两杯冷粥而已。当然，除了冷粥以外，让玛格丽特在比赛中走到最后并最终得到乔治国王和伊丽莎白女王接见的原因还有很多。如果这个食谱是比赛中的优胜作品的话，那些首轮就被淘汰的食谱又会是什么样子的呢？

FURTHER USES FOR COLD PORRIDGE

SURPRISE CREAM WITH OAT CRISPS. *Ingredients – Cream: 2 teacups thick cold porridge, ½ tin evaporated milk, ¼ oz. gelatine, 2 tablespoons castor sugar, vanilla essence, colouring if liked, 2 oz. sultanas. Crisps: 1 oz. flour, 1 oz. oatmeal, ¼ oz. lard, 1 teaspoon castor sugar, pinch of cream of tartar, ¾ level teaspoon baking powder, pinch salt, 1 tablespoon milk. Method: Whip [evaporated] milk till thick, gradually whip in the cold porridge. Dissolve gelatine in a little hot water, add to mixture with castor sugar, vanilla essence and colouring. Whip well and pour into glass dish. When set, decorate with raisins. Serve with sweetened oat crisps. Crisps: Mix dry ingredients and mix with milk, to a stiff consistency. Turn out and stamp with an apple corer. Cut into lengths and place on a greased tin. Bake in a moderately hot oven till crisp and lightly browned.*

Dainty Dishes for the Queen: Recipes of Oatmeal and Potato Dishes used by Girls in the Scottish Primary Products Contest, Scottish Education Department, 1944